Marina Paglione Ramia

Functional and structural studies of an atypical fungal endoglucanase

Marina Paglione Ramia

Functional and structural studies of an atypical fungal endoglucanase

Functional mechanism and biomolecular structure

ScienciaScripts

Imprint

Cover image: www.ingimage.com

This book is a translation from the original published under ISBN 978-3-330-77062-1.

Publisher:
Sciencia Scripts
is a trademark of
Dodo Books Indian Ocean Ltd. and OmniScriptum S.R.L publishing group

120 High Road, East Finchley, London, N2 9ED, United Kingdom
Str. Armeneasca 28/1, office 1, Chisinau MD-2012, Republic of Moldova, Europe
Managing Directors: Ieva Konstantinova, Victoria Ursu
info@omniscriptum.com

Printed at: see last page
ISBN: 978-620-8-63347-9

Summary

ACKNOWLEDGMENTS

First of all, I thank God and the beautiful opportunity of life.

To my parents Màrcia and Jûnior, for all their effort, love and dedication, always pointing the way and directing my choices until I could make them on my own.

To my brother Miguel for his companionship, complicity, support and friendly shoulder.

To Elder, for being my partner and for the life we are building together. I'm especially grateful for the maturity, strength and courage that drive me to overcome obstacles day after day.

To Andre Godoy, friend and fundamental collaborator in this work. Thank you for the laughs, parties, açals and for always being ready to help with affection and dedication.

To Prof. Igor for his guidance and opportunity.

To Prof. Alessandro and Prof. Joâo for always being willing to help.

To Dr. César Camilo and Dr. Marco Kadowaki for their patience, assistance and willingness to teach and help whenever I needed it.

To Prof. Nei Pereira Júnior for welcoming me so well to his wonderful city. To everyone I met there, for their pleasant welcome and willingness to help me.

To Dr. Roberto Maeda, who welcomed me, guided me and became a great friend.

To FAPESP for funding and providing all the support needed to carry out this project.

To the IFSC for all the infrastructure and support offered.

To Dr. Amanda for taking the time to read this thesis and for always being so helpful. To Vanessa, Aline, Grazi, Mel and Renata for your love and friendship, for our lunches and for the unconditional support I receive from you. Thank you so much!

To my college friends Gabi, Dani, Jéssica, Flàvia and Carla who, despite being far away, were always there for us in some way, even if it was just through a simple message.

To my soul brothers Galeno, Eric, Pri, Ivan and Wil. Because just seeing you made me overflow with joy. I'm grateful for our few but intense encounters. You make me believe in a better world. *Thank you!*

To all my friends in room 8 and the like. Thank you Danilo, Lis, Bruno, Caio, Atilio, Evandrinho, Vasia, Evandro, Karina, Mari and Carol.

To the technicians at the molecular biotechnology laboratory, Joâo, Josimar and Maria. My special thanks go to Livia, for always being willing to help, for her understanding and affection. Thank you for making our work easier.

To my family in general, Aunt Rose, Aunt Nê, Aunt Ana, Aunt Silvia, Uncle Fer, Uncle Du, my grandparents, great-grandparents, in short, everyone who contributed to me being here and now.

To my cousins Livia and Aline, who always gave me strength and support, and even though they didn't understand, made themselves available to read my work.

To Maria Avelar, dear master, and to yoga, for showing me that knowledge is not only acquired through the intellect.

Finally, I would like to thank everyone who directly or indirectly contributed to the completion of this work!

Thank you so much!

There are two ways of living life: one is to believe that there are no miracles;

the other is to believe that everything is a miracle."

Albert Einstein

SUMMARY

RAMIA, M. P. **Functional and structural studies of an endoglucanase from *Phanerochaete chrysosporium* of the glycoside hydrolase family 45**. 2015. 137p.

The importance of studying cellulases is not limited to the acquisition of scientific knowledge, but also to the great biotechnological potential they represent. This is because cellulose is the most abundant molecule present in nature and provides a wide range of sustainable products and processes. Many cellulase families have already been well characterized, while others remain unknown. Among the latter, the 45 family of glycoside hydrolases is the least characterized family of fungal cellulases, both structurally and functionally. It has recently been proposed to divide this family into three subfamilies and, so far, only members of subfamily A have had their enzymes structurally elucidated. In this work we report the crystallographic structure of the recombinant endoglucanase protein from *Phanerochaete chrysosporium* (PcCel45A), the first of the glycoside hydrolases in subfamily C, and its complex with cellobiose at 1.4 Â and 1.7 Â resolution, respectively. PcCel45A is a single-domain enzyme with a β-barrel structure and its general packaging resembles that of an anchor. The enzyme's active site forms a long groove on the surface of the structure, its catalytic center is different from the other published enzymes in this family and the catalytic aspartate, which acts as a proton acceptor in the inversion reaction, (AsplO) is not conserved. In addition, the crystallographic structure of this enzyme shows more similarities with β-expansins (plant proteins) and lytic transglycosylases (proteins that cleave peptidoglycan from bacteria) than with the other representatives of the 45 family, which makes it even more unique. In order to better understand how it works, site-directed mutations were made in the main residues of the active site. Asp121, known for participating in the inversion reaction of the other enzymes in the family as a proton donor, proved to be essential for the enzyme's activity, while other conserved residues such as Tyr25, Trp161 and Asp92 affected, but did not annihilate, the enzyme's activity, showing approximately 20%, 50% and 10% of the native enzyme's activity, respectively.

Keywords: Cellulase. Endoglucanase. Glycoside hydrolases. Crystallography of macromolecules.

Chapter 1

Introduction

"Every great advance in science has resulted from a new audacity of imagination. "

John Dewey

1 Introduction

Every day, plants produce approximately 50 kg of cellulose per person, which makes lignocellulosic biomass the most abundant polymer on the planet.[1-2] Because of this, its conversion into other products and bioprocesses is a sustainable proposal and has become the target of worldwide research.

Cellulose is a linear biopolymer made up of approximately 10,000 to 15,000 glucose molecules linked by β-1,4 glycosidic bonds. The cellulose molecule contains many polar hydroxyl groups, which allows the interaction between molecules of adjacent chains, forming fibers. These fibers are structurally strong and resistant to chemical and physical attack, so cellulose products are widely used in construction and paper production.[3] It is also used in the textile industry, such as cotton fibers, in the food industry, as additives, in the production of biofuels, organic acids, enzymes, biosorbents, among others.[4-5]

For some products, such as biofuels, cellulose needs to be hydrolyzed to fermentable sugars, which occurs through two main routes: acid or enzymatic hydrolysis.

Acid hydrolysis requires harsher conditions to achieve high levels of conversion to glucose, as well as forming fermentation inhibitors, which makes subsequent processes more difficult.[6] For this reason, large investments are being made to improve enzymatic hydrolysis. Its advantage over the first one is the high specificity of the enzyme for the substrate, with high yields, low energy costs and milder production conditions than chemical processes.[7]

Despite the great potential of enzymatic hydrolysis, some obstacles still need to be overcome. The recalcitrance of the cell wall, for example, hinders the accessibility of the enzymes to the substrate.[8] In addition, it is necessary to select more stable and efficient enzymes, in order to reduce the costs of the process. Because of this potential, many efforts have been made to facilitate this cellulose hydrolysis process and, as a result, much has been discovered about glycoside hydrolases.[9]

Glycoside hydrolases are enzymes capable of catalyzing the hydrolysis of glycosidic bonds between glucose polymers or between a glucose and another non-polysaccharide molecule.[10] The nomenclature of this group is due to the name of the substrate that these enzymes use, and may in some cases derive from their molecular mechanisms.

Based on the similarities of the amino acid sequences, a classification has been proposed that aims to better reflect the structural characteristics of these enzymes.[11] By comparing the sequences of glycoside hydrolases and other related enzymes, it has been possible to group them into families, which facilitates understanding of the evolutionary bases of this group, as well as a more complete organization of the information. The database that contains this classification and information is CAZy.[10]

Although many families already have their mechanisms very well established, with a wide variety of structures elucidated, there are still many gaps to be clarified. One of the families that little is known about is family 45. There is little structural information on this family and the information that does exist has led to the division of this family into three subfamilies, called A, B and C.[12] Although scarce, much of the information that exists on this family fits for the proteins belonging to subfamily A, which makes the search for functional information and the elucidation of the structures of the other subfamilies even more indispensable. With this in mind, this study aimed to elucidate the structure of an endoglucanase V belonging to subfamily C and to discuss this structure in comparison with the other enzymes in the family.

The microorganism that produces this endoglucanase is the *basidiomycete* fungus *Phanerochaete chrysosporium.* This fungus is classified as a *white-rot* fungus due to its ability to selectively degrade lignin.[13] In addition, it has a similar and equally advanced cellulolytic machinery to that of the fungus *Trichoderma reesei*, which is considered a model because it is the best studied.[14-15]

Chapter 2

Literature review

"SOY THE change you want to see in the world. "

Mahatma Gandhi

2 Literature review

2.1 Cell wall

The cell wall is a structure found in prokaryotes, fungi, plants and algae. Among these different groups, it differs in its chemical composition and microscopic structure. Its function in the plant cell is to provide mechanical strength so that the plant can grow in height and develop. It acts as an exoskeleton that controls the shape of the cell and allows the plant to withstand high turgor pressures.[16] By surrounding each cell individually, it acts to prevent sliding between adjacent cells, and at the same time allows the exchange of solutes and signaling molecules through structures called plasmodesmata.[17] It also acts as a diffusion barrier, limiting the size of the macromolecule that reaches the plasma membrane, preventing pathogens from entering.[16; 18]

The cell wall can be classified into primary and secondary cell walls. Primary walls are formed when cells are growing and are unspecialized and similar in all cell types. Secondary walls, on the other hand, are formed when cell growth has ceased and are usually specialized in their structure and composition, reflecting the differentiated state of the cell.[16]

2.1.1 Cell wall components

Plants convert solar energy into chemical energy stored in the bonds between sugar molecules, thus synthesizing carbohydrates that will serve as a source of energy and as protective blocks that make up the plant wall.[19] In the primary cell wall, cellulose microfibrils are impregnated in a polysaccharide matrix. The combination of these two structures gives the cell flexibility and mechanical strength at the same time. This matrix comprises two large groups of polysaccharides: hemicelluloses and pectin, as well as a small amount of structural proteins. Depending on the type of cell and plant species, the polymers in the matrix vary. The diversity of the polymers that make up the plant cell wall can be seen in Table 1.

Table 1 - Structural components of the plant cell wall.

Class	Examples
Cellulose	Microfibrils of (1,4)β-D-glucan
Polysaccharide matrix:	Homogalacturonan
Pectins	Rhamnogalacturonan
Hemicelluloses	Arabinan
	Galactan
	Xyloglucan
	Xylan
	Glucomannan
	Arabinoxylan
	(1,3) β-D-glucan
	(1,3-1,4) β-D-glucan (grasses only)
Lignin	Coniferyl alcohol
	Sinapil alcohol
	P-coumaryl alcohol

Structural proteins	HRGPs - GRPs, PRPs
	Extensins
	Arabinogalactan proteins (AGPs)

Source: Prepared by the author

2.1.2 Carbohydrate structure bonds

The anomeric carbon is always linked to the hydroxyl of another sugar or hydroxylamine (NH_2-OH). The bond between D-glucose and another sugar can occur on the oxygens of carbons 2, 3, 4 or 6, with the oxygen of carbon 5 being unavailable because it forms the ring.

Polysaccharides are named according to the main sugars that make them up. They all have a sugar skeleton which is designated by its last name. For example, like cellulose, xyloglucan is composed of a skeleton of (1,4)β-D-glucose residues. However, it has side groups containing xylose, galactose, and often a terminal fucose, as seen in figure 1.

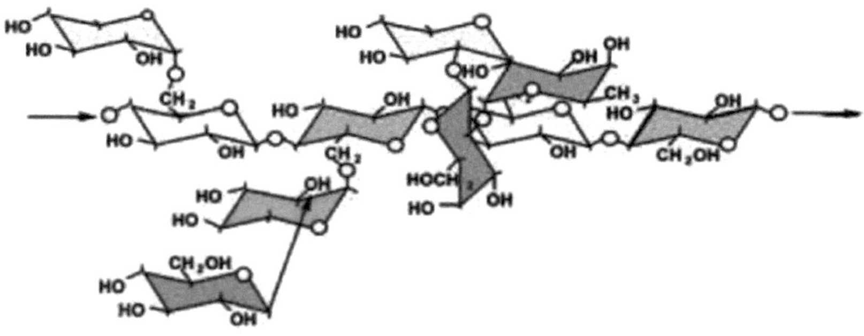

Figure 1 - Molecular structure of xyloglucan with four glucose units (in white and light orange) joined by β1,4 bonds, with α-D-xylose units linked to them (blue). About half of these units contain β-D-galactose and L-fucose residue extensions (dark orange and violet respectively).

Source: CARPITA et al.[20]

2.1.3 Polysaccharide matrix

Also known as hemicellulose, the polysaccharide matrix is made up of a heterogeneous group of polysaccharides. This matrix has an inferior structural organization, mainly due to the branching and non-linear structure of the polysaccharides that make it up. The variation in its composition depends on the type of tissue and the species of plant. For example, in the primary walls of dicotyledons, the most abundant polysaccharide is xyloglucan. The side groups of xyloglucan do not allow it to group together into a crystalline microfibril (figure 1). Because they are larger (50 to 500 nm) than the space between the cellulose microfibrils (20-40 nm), they act to bind the microfibrils together. The polysaccharide matrix bonded with cellulose intertwines with lignin molecules, forming a large network of interactions between the atoms, which provides the observed structural strength.[21]

The hemicellulose fraction can also contain other important polysaccharides such as glucuronoarabinoxylan and glucomannans. Secondary walls usually contain less xyloglucans and more xylans and glucomannans. The cell wall of grasses contains only a small amount of xyloglucan and pectin, which have been replaced by glucuronoarabinoxylan and (1,3-1,4) β-D-glucans. The latter are the same as cellulose except that they have single β1,3-D-glucosidic bonds between multiple β1,4-D-glucosidic bonds, thus introducing kinks in the cellulose chain.[17]

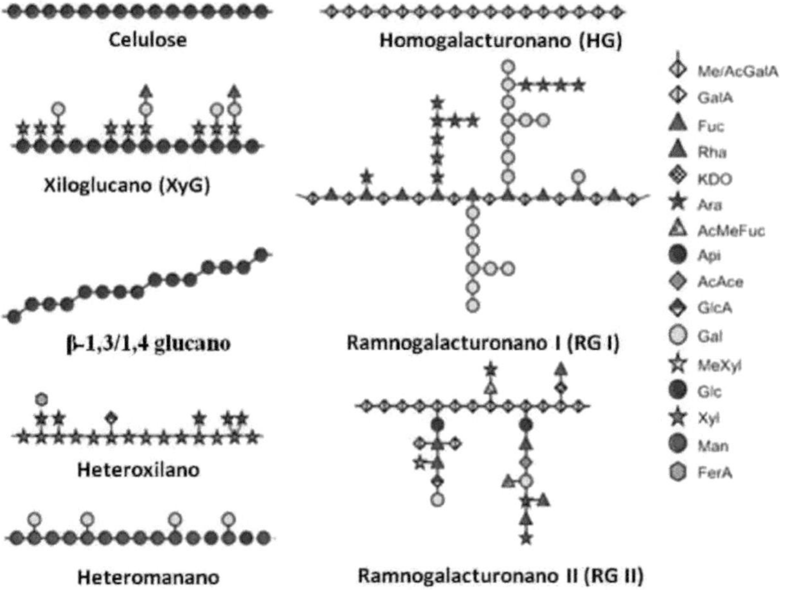

Figure 2 - Schematic representation of the polysaccharide structures found in the cell wall. Cellulose, xyloglucan (XyG) and (1,3ι1,4)-β-D-glucan have a skeleton based on glucose (Glc), heteroxylan on xylose (Xyl), heteromannan on mannose (Man), homogalacturonan (HG) and rhamnogalactorunan II (RG I) on galacturonic acid (GalpA) and RG I alternated with the disaccharide of (GalpA) and rhamnose (Rhap).

Source: Adapted from DOBLIN et al[17]

2.1.3.1 Pectin

Pectin comprises one of the most complex families of polysaccharides, as it can be made up of up to 17 different types of monosaccharides, which form more than 20 types of bonds.[22]

They are a group of heteropolysaccharides that normally have acidic sugars such as galacturonic acid

(GalpA) and neutral sugars such as rhamnose, galactose and arabinose.

Homogalacturonan (HG) is the most abundant pectic polysaccharide in the cell wall. It is made up of a linear chain of GalpA residues linked by an α-1,4 bond (figure 2). These residues are sometimes methyl-esterified at C6, as well as O-acetylated at C3 and/or C2.[23]

Other heteropolysaccharides in this group are considerably more complex and include rhamnogalacturonan I and II (RG), xylogalacturonan, arabinan and arabinogalactan I and II. Rhamnogalacturonan I consists of an interspersed GalpA and rhamnose skeleton, with α-arabinan and β-galactan side-chains, some of which have terminal α-fucoses. Rhamnogalacturonan II, on the other hand, is the most complex heteropolysaccharide, as it is made up of a HG skeleton with at least 8 GalpA residues with side chains of 12 different types of sugar in more than 20 different types of bonds.[23]

Pectins constitute the most soluble fraction of wall polysaccharides and often form a gel.[16] The pectin gel is formed when HG portions interlock to form a three-dimensional crystalline network in which water molecules and solutes are trapped.[24]

2.1.3.2 Structural proteins

The cell wall has a variety of classes of structural proteins, which are important for the functioning and structure of the cell wall. They act in the assembly and remodeling of the wall when it is growing and developing. These proteins are classified according to their predominant amino acid composition. The most abundant structural proteins of the cell wall are known as hydroxyproline-rich glycoproteins (HRGPs) which include glycine-rich proteins (GRPs), proline-rich proteins (PRPs), extensins and arabinogalactan proteins (AGPs).[17]

2.1.3.3 Lignin

Lignin is a three-dimensional polymer made up of highly branched phenylpropanoid units and comes from three precursor phenolic compounds called monolignols: the coniferyl, sinapyl *and coumaryl* alcohols (Figure 3).[25]

The monolignols are directed to distinct regions of various types of cell wall (Figure 3 b) where they are polymerized into a structure with distinct biophysical characteristics, which offers rigidity and resistance to compression of the cell wall.[21;26]

The proportion of these three monomers varies between species, plant organs and even between the layers of the same cell wall.[27] However, it is most present in the secondary wall and can also occur in the primary wall and middle lamella, along with cellulose and hemicellulose.

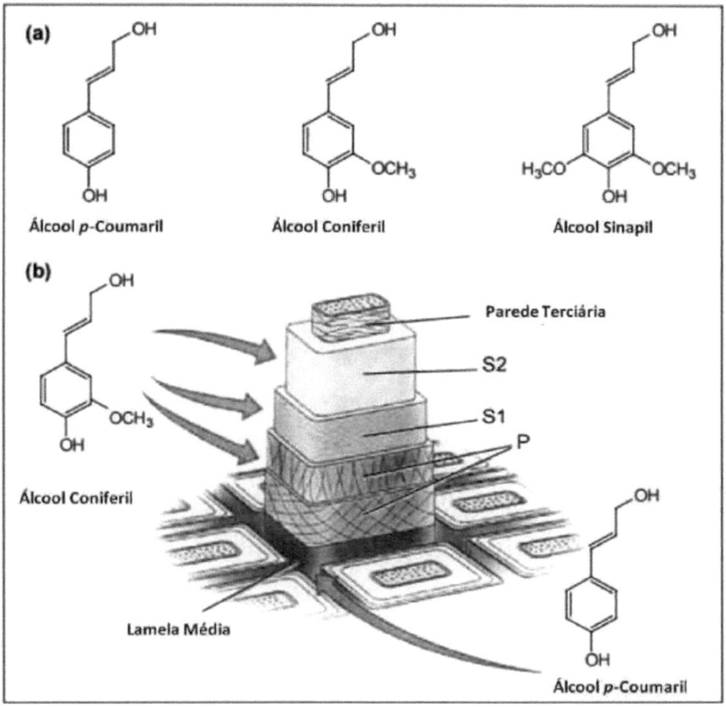

Figure 3 - Monolignols and the different concentration targets in the cell wall. A) Chemical structures of monolignols. B) Representation of the xylem of a conifer. The *p-coumaryl* alcohol is deposited preferentially in the middle lamella and the coniferyl alcohol in the secondary wall.

Source: Adapted from DAVIN et al.[28]

Unlike cellulose, starch and other polymers, which are linked in a simple, repetitive way, lignin polymers are more complex. As there are multiple C-C and C-O-C bonds in each monolignol unit, there are several branches in all three dimensions.

The association of lignin with polysaccharides in the matrix is the factor that most contributes to the recalcitrance of the cell wall.

2.1.3.4 Cellulose

The mechanical strength of the cell wall is provided by cellulose microfibrils. These microfibrils are made up of glucose monomers, which are joined together by β1,4 glycosidic bonds. The glucans that make up the microfibril are closely aligned and bonded to each other, forming a highly crystalline structure that excludes water and is therefore inaccessible to enzymatic attack.

The precise structure of cellulose is still uncertain, but current models suggest that it consists of crystalline domains linked to less organized regions, called "amorphous". In the crystalline domains, glucans of the same chain are joined by covalent bonds, while adjacent glucans (interchains) are interacting by non-covalent bonds, such as hydrogen bonds and hydrophobic interactions.[(27;) (29)] These inter- and intra-chain interactions are responsible for the formation of the fibril and crystallinity of cellulose.

A complete model of cellulose degradation can be obtained by measuring three parameters: the degree of polymerization (GP), degree of oxidation and crystallinity.[30] All these parameters vary depending on the source and age of the tissue.[31]

2.2 Plant growth

For plants to grow, the cell wall, which is rigid, somehow needs to loosen to make room for cell growth. The type of wall loosening involved in plant cell expansion is called relaxation stress. This term is used in a mechanical context as the force exerted per unit area. Plant cells accumulate osmotically active solutes that attract water into the cell by osmosis. As the cell is limited by the cell wall, there is an increase in turgor pressure and in the opposing force, which is the tensile stress inside the cell wall, as can be seen in Figure 4.[19]

It is not yet well described how the opening of the cell wall occurs, but there are some hypotheses. According to the proposition of acid relaxation due to the action of auxin, one mechanism that would cause the relaxation stress would be the acidification of the wall, resulting from the extrusion of protons through the plasma membrane. This would explain why cell wall loosening increases in the presence of acidic pHs.32

Because it is composed of a hydrated polymer, plant cell walls have physical properties that are intermediate between those of solids and liquids. These properties are called rheological. The stress caused to cell walls is the result of an increase in turgor pressure which, consequently, causes a change in the cell's rheological properties. [33] The figure below shows a model of how the cell wall loosening process can occur.

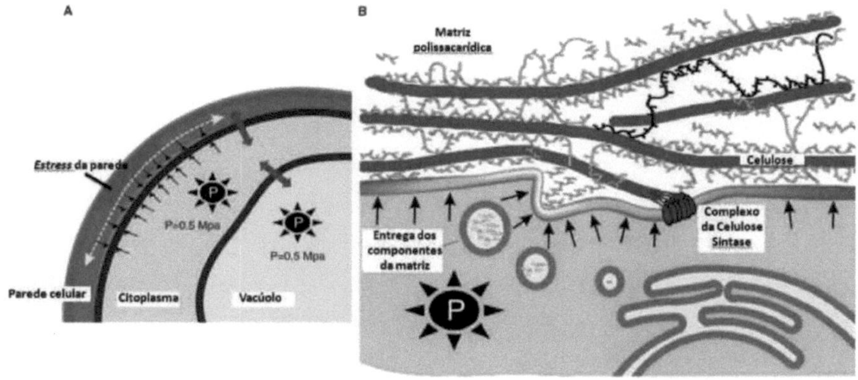

Figure 4 - A) Turgor pressure (P) is the force against the plasma membrane. It is directed outwards and is practically uniform and isotropic within all the internal compartments of the cell. It is counterbalanced by an equal and opposite inward force exerted by the cell wall against the plasma membrane (black arrows). This inward force generates a tensile stress in the plane of the cell wall (yellow arrow). Turgidity pressure arises from the accumulation of osmotically active solutes in all intracellular compartments, resulting in iransmembrane water flows (red arrows) to establish osmotic balance. The swelling of the protoplast is limited by the cell wall, resulting in a turgidity pressure inside the cell and a stress in the plane of the cell wall. B) A sketch of cellulose microfibrils (brown bars) being "spun" from the cellulose synthase complex (red) and linked together by matrix polysaccharides (gray). Newly synthesized wall materials are deposited in a relaxed state and become a support only after they integrate into the wall network. The widening of the cell wall makes them elastically stretched so that they begin to create a resistance. At this point, biophysical loosening of the cell wall network by expansins or other loosening agents is necessary to relieve wall tension, allow water absorption and continue cell growth.

Source: Adapted from COSGROVE et al.[19]

In 1989, in order to determine whether cell wall stretching was caused by a simple flow of viscoelasticity or whether there was the action of proteins, Cosgrove [34] carried out a series of experiments which confirmed the action of proteins in loosening the cell wall. McQueen[35] then managed to isolate two of these proteins and observed that when they were placed in small quantities in heat-denatured cell walls, wall extension occurred again. These proteins were named expansins. Despite the action of expansins in wall induction, it was found that they had no hydrolytic activity and their mechanism of action is still uncertain.

Many studies have shown that expansins are regulators of cell growth. It is known that expansin genes are expressed at the right time and place to control wall growth and that reducing the expression of these genes inhibits wall growth.[36-37] These proteins also induce pH-dependent extension of isolated cells, and their application to living cells stimulates cell wall growth.

2.3 The hydrolysis of cellulose

2.3.1 Glycoside hydrolases (GH)

Glycoside hydrolases form an extensive group of enzymes that catalyze the cleavage of the O-glycosidic bond.[38] They are found in almost all groups of organisms, the only exceptions being archaea and some eukaryotic parasites.[39] Considering the wide variety of monosaccharides, the diversity of inter-saccharide bonds and the fact that almost all molecules can be glycosylated (including proteins, nucleic acids, lipids, antibiotics), it is possible to imagine the wide variety of enzymes required for specificity in glycoconjugates, oligos and polysaccharides.[10]

Since the variety of carbohydrates greatly exceeds the number of possible protein packages, glycoside hydrolases have evolved from a limited number of progenitors, diversifying their function and acquiring new specificities.[10]

Approximately 26 years ago, Henrissat and colleagues used a tool called Hydrophobic Cluster Analysis (HCA) to compare 21 amino acid sequences of cellulases available at the time. They found evidence of a series of homologies between fungal and bacterial cellulases, which allowed them to classify these enzymes initially into six families ranging from A to F.[40]

Since then, the number of glycoside hydrolase genes sequenced has increased exponentially, followed by an increase (albeit smaller) in the number of crystallographic structures elucidated.[41] This has made a more rigorous classification necessary in order to avoid ambiguities and provide a single classification.

The IUB enzyme nomenclature (1981) indicates the type of reaction that the enzyme catalyzes and its substrate specificity. Thus, glycoside hydrolases are indicated by the code EC 3.2.1.x, with the first three digits indicating that they cleave O-glycosidic bonds and the last number reflecting the main substrate and, in some cases (e.g. amylases EC 3.2.1.1 and EC 3.2.1.2) the mechanism of the reaction.[38]

As these enzymes are classified according to the similarity of their amino acid sequences, this classification correlates the enzymatic mechanism and protein packaging and not the specificity for the substrate.[(10)] Therefore, in the same family we can find enzymes with different specificities for substrates, and enzymes that hydrolyze the same substrate are also found in different families.[42]

The hydrolysis of the glycosidic bond is catalyzed by two amino acid residues of the enzyme: a general acid (proton donor) and a nucleophile (base).[39] Often these residues are a glutamic and/or aspartic acid due to their carboxyl group. Depending on the spatial position of these catalytic residues, hydrolysis occurs through total retention or general inversion of the anomeric configuration.[43] The distance between the carboxyl groups of these residues is generally 5-6 Â for retention and 9-10 Â for inversion enzymes. In the latter case, this greater distance is due to the need for a water molecule in the enzyme's active site during the hydrolysis reaction.[44]

2.3.2 Transglycosylases

There are currently 135 known families of glycoside hydrolases (October 2015), which include not only glycosidases, but also transglycosylase enzymes (EC 2.4.1). Glycosidase enzymes often exhibit a wide range of substrate specificity and are often capable of transferring glucose residues between substrate molecules by transglycosylating glycosidic bonds [38,10,45].

They are classified in the GH2, GH13, GH16, GH23, GH31, GH70, GH77, GH102, GH103 and GH104 families (CAZy, October 2015).

The GH102 family is made up of lithium transglycosylases, the prototype of which is the *Escherichia coli* lithium transglycosylase enzyme (EcMltA).[46]

This enzyme is responsible for cleaving the bacterial cell wall, thus maintaining a recycling of peptidoglycans, which facilitates the growth and division of the bacteria, as it allows a local opening of the wall without breaking its integrity. It has two domains, with domain I being structurally similar to that of the GH45 family, containing the double-Ψ-β-barrel motif. The lytic activity of these enzymes occurs at the β1,4 glycosidic bonds between N-acetylmuramic acid (MurNAc) and N-acetyl-glucosamine (GlcNAc), the two sugar units that form the peptidoglycan strand.

The reaction catalyzed by this enzyme differs partially from most glycoside hydrolases in that a single residue, usually an aspartic acid (Asp308), acts as an acid-base, causing the reaction to take place in two steps. In the first, this aspartate acts as an acid, donating a proton to the oxygen of the glycosidic bond. The cleavage of the glycosidic bond leads to the formation of a positively charged oxocarbenium ion intermediate. In the second step of the reaction, the same residue, in its ionized form, acts as a base, abstracting a proton from the hydroxyl group of C6, activating O6 for an intraresidue nucleophilic attack on the C1 atom of the oxocarbenium ion. This completes the reaction and results in the formation of the product 1,6-anhydroMurNAc.[46] A schematic of the reaction can be seen in figure 5.

Figure 5 - Proposed reaction mechanism for the enzyme Iransglycosylase from *E. coli* GH102. In this mechanism, an aspartate donates a proton to the oxygen of the glycosidic bond, forming a positively charged intermediate. The same residue then abstracts the proton from C6, activating O6 for an intramolecular attack.

Source: VAN STRAATEN et al.[46]

2.3.3 Cellulases

Cellulases differ from other glycoside hydrolases in that they cleave β1,4 bonds between glucose residues.[2] The classic model of enzymes that cleave cellulose resulting in glucose monomers classifies three types of enzymatic activity, which are: endo-1,4- β-glucanases (EC 3.2.1.4), exo-1,4-β-glucanases and cellobiohydrolases (EC 3.2.1.74 and EC 3.2.1.91 respectively) and cellobiases or β-glucosidases (EC 3.2.1.21). They all act synergistically by cleaving the β1,4 glycosidic bonds that connect glucose residues, differing only in the mode of catalysis.[47]

In addition to the enzymes described above, some fungi also secrete an extracellular enzyme called cellobiose dehydrogenase (CDH; EC 1.1.99.18). This enzyme has two domains, one of which contains a flavin adenine dinucleotide (FAD) and the other a heme group. It acts on cellobiose and cello-oligosaccharides, which are oxidized to their corresponding lactones during the reductive half-reaction of the FAD cofactor and subsequent hydrolysis to aldonic acids.[(48-) (49)]

In addition, new types of enzymes that were originally called carbohydrate-binding modules (CBM 33) have been identified as acting in the cleavage of cellulose. These enzymes, now called lithiated polysaccharide monooxygenases (LPMOs), depend on copper AA10 and GH 61 (now called AA9) and act like oxidative enzymes. They act together with cellobiose dehydrogenases by adding breaks to the structure of cellulose and hemicellulose, increasing the hydrolytic activity of cellulases and hemicellulases.[50]

Currently, there are many studies on the structural and functional characterization of cellulases.[51] It is known that most of them are composed of a catalytic domain, responsible for enzymatic hydrolysis; a carbohydrate-binding module (CBM); and a binding region rich in threonine, serine and proline residues by which the domains are connected.[52]

2.3.3.1 Endoglucanases

Endoglucanases (EC 3.2.1.4) attack the amorphous part of cellulose by cleaving the internal bonds in the microfibrils, reducing the size of the chain, generating oligosaccharides of different sizes and new free ends. Thus, they act by making new sites available for the action of cellobiohydrolases. They seem to have little capacity to degrade crystalline cellulose, due to the strong hydrogen bonds that occur in these fibers, making it inaccessible for these enzymes to come into contact.[47]

This class of enzymes is distributed in 13 families (CAZy, October 2015) belonging to eukaryotic organisms, bacteria, fungi and archaea.[53] Generally, its active site consists of an open groove, or slit-shaped, which allows the linear cellulose chain to accommodate randomly.[54] In order to effectively bind to the enzyme's active site, the substrate molecule interacts with multiple subsites (approximately 4 to 7) via hydrogen bonds and/or stacking of aromatic residues with amino acids in the active site.[55]

To carry out hydrolysis, endoglucanases keep the glucose units of the -1 and +1 subsites in the chair conformation while "stretching" the glycosidic bond[56] or "bending" the glycosidic bond, bringing it into an axial position, while distorting the -1 glycosyl unit into a boat conformation and replacing the +1 sugar.[57]

2.3.3.2 Exoglucanases

This group includes cellobiohydrolases (CBH) and glucanohydrolases (GH). Glucanohydrolases (EC 3.2.1.74) are capable of releasing glucose directly from the cellulose fiber.[2]

CBHs (EC 3.2.1. 91) can be divided into two types: CBH1, which hydrolyze the reducing ends, and CBH2, which hydrolyze the non-reducing ends. Both act on crystalline cellulose, resulting in a slow and gradual reduction in its degree of polymerization. Their sites on are generally in the shape of tunnels.[9]

2.3.3.3 β-glucosidase

The β-glucosidases (EC 3. 2.1. 21) act by breaking the β1,4 bond, when the carboxyl group of the conserved residue of the active site, usually a glutamic acid, donates a proton to the bond and the carbon ion formed is stabilized by the aspartic acid residue. These enzymes act on cellodextrin residues (GP < 7) and cellobiose, hydrolyzing them to glucose.[2] Similarly, exoglucanases and β-glucosidases seem to be inhibited by their own hydrolysis product.

2.3.3.4 CBM

As described above, many cellulases have this domain linked to the catalytic domain. It is known that CBMs help the enzyme to effectively recognize the surface of crystalline cellulose, bringing the active site into contact with the substrate and ensuring its correct orientation. This domain contains between 30 and 200 amino acids and there can be more than one in each enzyme.[51; 58]

2.3.4 Mode of action and synergism

The classic view of cellulose hydrolysis states that the endoglucanase enzymes initially hydrolyze the internal bonds of the cellulose fiber at random. As a result, they produce small cuts, releasing the reducing and non-reducing ends of the substrate. The exoglucanases then act on these ends, releasing cellulose oligosaccharides and cellobiose. The products of the endoglucanases and cellobiohydrolases are inhibitors of these enzymes, so *there is the* action of the β-glucosidases which will act on the cellobioses and release glucose.[59]

With the discovery of these oxidative enzymes (CDH, CBM33 and GH61), a new paradigm also emerged in relation to the enzymatic degradation of cellulose, with the latter facilitating the action of classic cellulases.[60] Little is understood about the action of these enzymes, but the most accepted model suggests that the LPMOs act on the crystalline cellulose fiber, producing partially amorphous regions with oxidized ends that function as "entrances" for the action of the GHs. The CDHs would work by providing electrons for the LPMOs.[61] A schematic representing this model can be seen in figure 6.

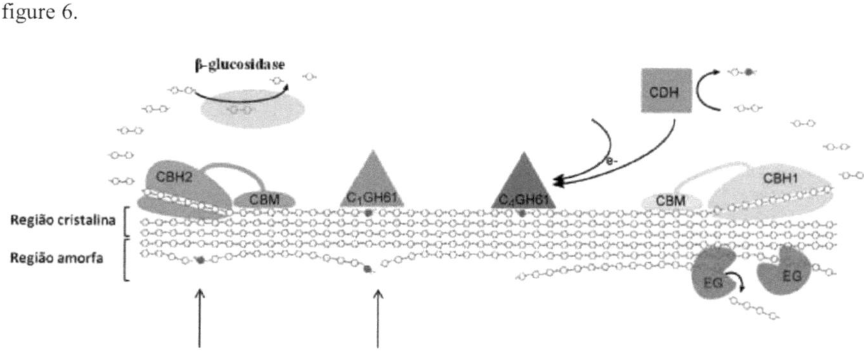

Figure 6 - Diagram of enzymatic degradation by the enzymes endoglucanases (EGs in orange), cellobiohydrolases (CBH1 in light blue, CBH2 in moss green), β-glucosidases (in lime green), cellobiose dehydrogenase (CDH in gray) and GH61, which oxidizes C1 (in pink), facilitating the action of CBH2, and C4 (in purple), facilitating the action of CBHl (reducing end). Oxidized sugars are shown in red. The arrows show new attack points for CBHs from the middle of the cellulose chain.

Source: Adapted from HORN et al.[60]

2.4 GH45

The 45 family of glycoside hydrolases (also known as K-family cellulases) is basically made up of β1,4 and β1,3/1,4 glucanases that belong to a wide range of living beings such as bacteria, fungi, plants, protozoa and some animals such as molluscs, insects and nematodes.[62] They share a small catalytic domain (20 kDa) and act on a wide variety of substrates such as cellulose and its derivatives. Predictions of the catalytic residues allowed the GH45 family to be classified as enzymes that hydrolyze the substrate by inversion of the anomeric carbon.[63]

With the sequence comparison between GHs 45 and other enzyme classes, it was possible to observe that they are distantly related to plant expansins. Due to these phylogenetic differences, Igarashi proposed subdividing this family into 3 subfamilies called A, B and C, as shown in figure 7.[12]

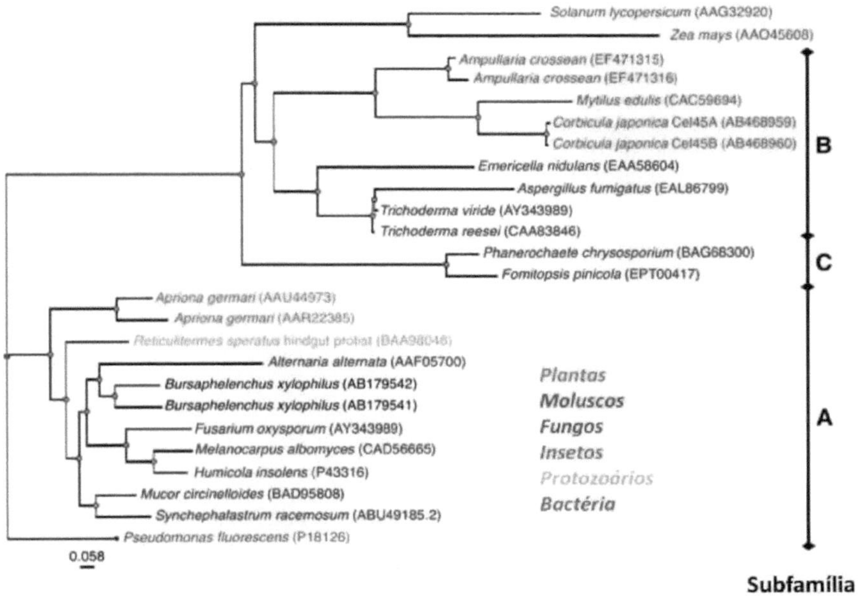

Figure 7 - Phylogenetic tree of the GH45 family. Sequences from plants, molluscs, fungi, insects, protozoa and bacteria have been included and colored according to the legend. The vertical bars on the right correspond to the subfamilies, according to the classification proposed by Igarashi.[12] The plant sequences are expansins, so they are not included in the subfamily classification.

Source: Adapted from PAYNE et al.[64]

2.4.1 Subfamily A

Subfamily A is the best studied and also the first to have a representative with a determined crystallographic structure.[63]

The general shape of the enzymes in this subfamily is that of a flattened sphere that has a structural motif known as the double-Ψ-β-barrel, formed by six parallel and antiparallel β-sheets that form two psi structures, as can be seen in Figure 8.[10]

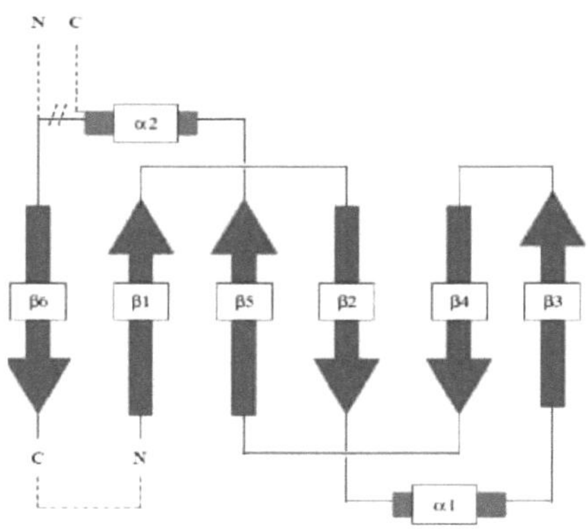

Figure 8 - Topological representation of the double-Ψ-β-barrel. The blue arrows indicate the β-sheets, while the red boxes indicate α-helices.

Source: CASTILLO et al.[65]

The first representative of this family to be studied is endoglucanase V from *Humicola insolens* (Hi_EGV). Structural data showed that this enzyme acts by inverting the anomeric carbon. This enzyme is composed of a catalytic domain in the N-terminal region, followed by a *linker* region attached to a CBM in the C-terminal region. Three sheets of the β-barrel domain form the active site. On the middle strand is the Asp121 residue, responsible for the protonation of the glycosidic oxygen. The other catalytic residue is Asp10, which acts by activating the nucleophile (water) for the deprotonation event and is located on the other side of the active site. They are approximately 8.0 Â apart, which allows this nucleonic water molecule to bind between them. As such, Asp10 acts as a catalytic base, activating a water molecule for nucleophilic attack. A schematic representation of the reaction can be seen in figure 9.

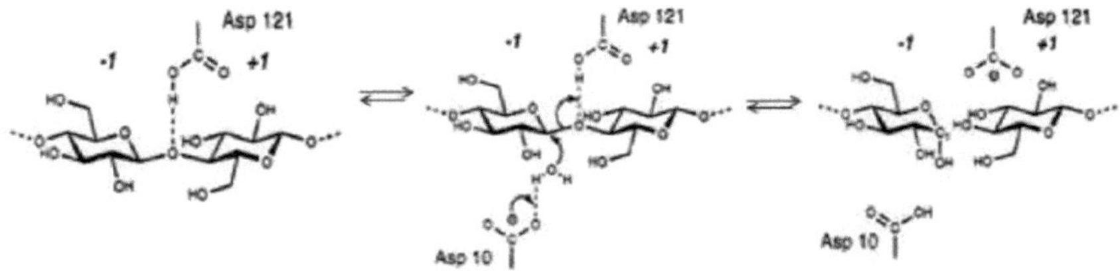

Figure 9 - Schematic representation of the hydrolysis mechanism of the Hi_EGV enzyme. This enzyme works by inverting the anomeric carbon. Asp121 donates a proton to the oxygen of the glycosidic bond. The base of the reaction represented by Asp10 acts by activating a water for nucleophilic attack.

Source: DAVIES et al[63]

Kinetic studies of Hi_EGV have shown that there are at least six energetically significant subsites for substrate binding.[66] Cleavage occurs between the -1 and +1 subsites, with the exit groups considered to be +1, +2, +3, +n. Hi_EGV has 4 subsites before the cleavage point, marked as -4 to -1 and two after the cleavage point, marked as +1 and +2 as observed in the structure of this enzyme with the celohexaose ligand (pdb: 4ENG). The substrate cleavage site can be seen in Figure 10.

This enzyme acts on the amorphous part of cellulose, and no activity has been identified on crystalline cellulose.

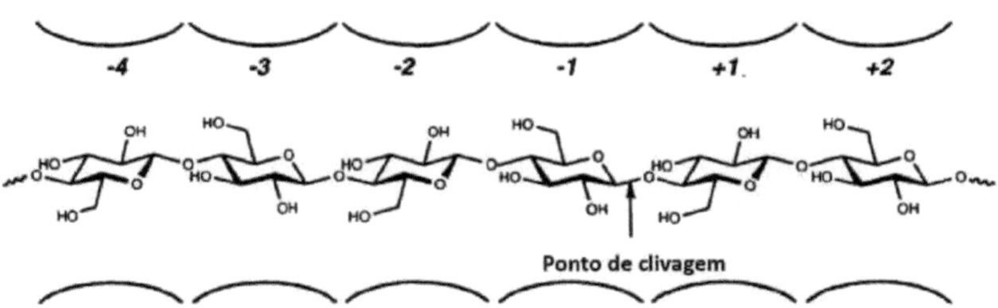

Figure 10 - Diagram showing the cleavage point of celohexaose by Hi_EGV. When this enzyme cleaves celohexaose, it preferentially releases cellobiose or cellotriose as a product which occupies the +1 and +2 subsites (+3 if the product is cellotriose).

Source: Adapted from DAVIES et al.[63]

Another structurally characterized enzyme that belongs to subfamily A is endoglucanase V from *Melanocarpus albomyces* (maEG). This enzyme is very similar to Hi_EGV with 75% sequence identity and RMSD less than 1 Â.[67]

Unlike Hi_EGV, this enzyme lacks the CBM and the linker region between the CBM and the catalytic domain. However, its overall shape is very similar to that enzyme. It also has a cleft similar to

Hi_EGV and the catalytic residues are conserved (Asp120 and Asp10 in maEG). It also showed greater activity on amorphous cellulose and is used in the textile industry to give jeans an aged appearance.

Its structure with cellobiose revealed sugar binding at the -2 and -3 subsites and, as with Hi_EGV, there are no aromatic residues on its surface. Binding to the substrate occurs by means of *loops* surrounding the sugar, allowing it to be correctly positioned in the active site.

Another enzyme with a determined crystallographic structure is the endoglucanase from *Humicola grisea*, which has more than 95% identity with Hi_EGV, with only 3 residues different from the latter and no publications on it.

1.1.2 Subfamily B

It has a solved crystallographic structure which is the endoglucanase from the mussel *Mytilus edulis* (Me_EGV). This enzyme consists of a catalytic module with 181 amino acids and has been solved to 1.2 Â resolution (pdb 1WC2).

It shows only 13% identity with Hi_EGV, reflecting the great difference between the subfamilies. However, the structural motif of the double-Ψ-β-barrel overlaps with an RMSD of 1.6 Â for 105 Cα atoms combined, and the structures are similar around the active site. The catalytic base (Asp24) and the catalytic acid (Asp132) are conserved. In both structures there is a conserved tryptophan (Trp64 in Me_EGV and Trp18 in Hi_EGV) in the -4 subsite region.

1.1.3 Subfamily C

For this subfamily, no structural information was available, with only the biochemical characterization of the endoglucanase V enzyme from *Phanerochaete chrysosporium* having been carried out.[12]

2.5 The basidiomycete fungus *Phanerochaete chrysosporium*

Basidiomycetes are a phylum of fungi so called because they reproduce by forming a structure called a *basidium.*[68] Within this phylum there are two main classes, called brown and white rot fungi. The difference between them is that white rot fungi are able to degrade all the components of lignin, while brown rot fungi are only able to modify the lignin that is left as a residue in the decaying wood.[68] This difference occurs because brown rot fungi do not have a class of peroxidase enzymes.

The fungus *Phanerochaete chrysosporium* belongs to the class of white rot fungi and is part of this small group of fungi capable of degrading lignin.

As mentioned above, the lignin present in lignocellulose is the most difficult obstacle that cellulolytic enzymes find to cleave cellulose efficiently. Because of this, this fungus has been the target of research for the last 40 decades, being the best studied basidiomycete.[69,25]

In addition to lignin degradation, another factor that has contributed to the extensive study of this fungus is its ability to degrade a wide range of toxic environmental residues, due to its oxidative bioremediation potential and tolerance to toxic substances.[70-71]

Proteomic studies of *Phanerochaete chrysosporium* have shown that the production of cellulolytic enzymes by this fungus not only depends on the carbon source, but is also influenced by the particle size and complexity of the lignocellulosic biomass.[72]

It is known that this fungus encodes more than 240 enzymes active on carbohydrates, including 166 glycoside hydrolases , 14 carbohydrate esterases and 57 glycosyltransferases, while enzymes that degrade lignin include lignin peroxidase, manganase peroxidase and oxidases.[69]

The way in which it degrades lignin is through the production of oxidases and peroxidases that generate highly reactive and non-specific free radicals that bombard and react with lignin, generating a series of spontaneous cleavage reactions.[73-74]

In addition to lignin, this fungus is also one of the most powerful decomposers of cellulose and hemicellulose, as it grows on decaying wood, an environment rich in these polysaccharides.[69; 75]

2.6 Fungi as an expression system

The yeast *Saccharomyces cerevisiae* was the first fungus to be transformed with exogenous DNA. A strain of this yeast was constructed to contain two separate point mutations within a single locus, thus preventing any reversion of the gene to functionality. A bacterial plasmid containing a copy of the inactivated gene was introduced into this strain. The exogenous DNA was the source for restoring the activity of the missing gene, paving the way for the development of this technology, making it possible to manipulate the fungal genome for the heterologous production of proteins.[74]

Ultraviolet light, radiation or chemical reagents are used to create a genetically modified strain, which develops only after the plasmid has been inserted. To do this, colonies of conidiophores are exposed to one of these mutagens in a complex medium and transferred to a minimal medium. All those that fail to grow in this minimal medium will, in a second step, be tested in this same medium now containing an additive that should restore growth (leucine, tyrosine, alanine, for example). If growth occurs again with the addition of this additive, it means that the specific mutation has been identified.[74] A schematic of this technique can be seen in figure 11.

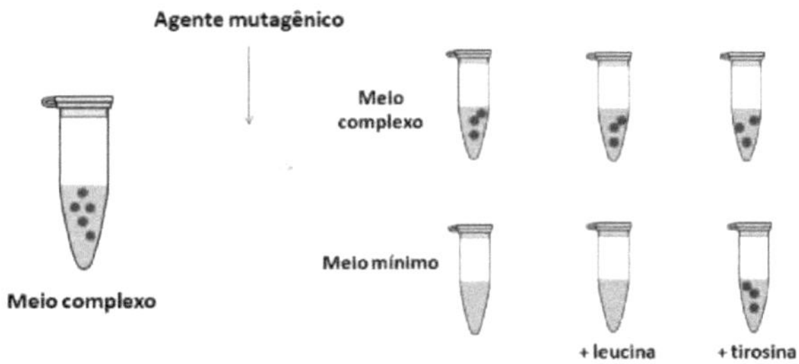

Figure 11 - Diagram of the production of mutant strains. Initially, the fungus is kept in a complex medium, followed by contact with a mutagenic agent. The strains are then inoculated into minimal medium, followed by the addition of an additive which would restore their growth.

Source: Prepared by the author.

Transformation in filamentous fungi is carried out using protoplasts or asexual conidia. To use the protoplasts, the cell wall is digested with hydrolytic enzymes, followed by the use of PEG (polyethylene glycol) and calcium to facilitate the entry of DNA by inducing membrane rupture. The incubation period that follows aims to regenerate the protoplast wall in an osmotically stable medium.

Filamentous fungi, such as *Apergillus nidulans*, have considerable potential as an expression system for heterologous proteins.[76] They are capable of continuous expression and their filamentous morphology allows the cells to be separated by filtration. In addition, the metabolic diversity of *Aspergillus* presents an opportunity to expand the range of possible heterologous metabolites in this host organism. *Aspergillus* is especially promising as a system for recombinant proteins, as it possesses the cellular machinery for protein translation, folding and post-translational modifications that bacteria do not, as well as the ease of cultivation and the export of the protein directly into the culture medium, which facilitates purification.[77]

Chapter 3

Objectives and justification

"An obstacle is what you see when you take your eyes off the goal you want to achieve. "

Bob Marley

3 Objectives

3.1 General objectives

The functional and structural study of cellulases helps to gain a better understanding of the molecular elements responsible for the enzymes' characteristics. It is known that the cost and low yield of these enzymes limit their applications in industry. Therefore, research aimed at linking structure and function can contribute to the production of more stable, active and specific enzymes.

From a scientific point of view, the study of enzyme classes with little information available, represents a vast field of possibilities, such as a greater understanding of enzyme mechanisms, as well as a greater understanding of the evolutionary relationships between the different enzyme classes.

In view of the above-mentioned importance, the aim of this study was to broaden our understanding of the GH45 class of enzymes by choosing to produce and characterize the endoglucanase V enzyme (PcCel45A) from the fungus *Phanerochaete chrysosporium.* Motivated by the information that this fungus produces a complex of Hgnocellulolytic enzymes and by the need to broaden our knowledge of the C subfamily of GH45, we aimed to elucidate the structure of this target. There were already some biochemical studies of this enzyme, produced by another expression system, and so we sought to compare the two and complement these studies. Subsequently, in order to gain a better understanding of the molecular basis of the enzyme's function and stability, mutagenesis was carried out on some specific residues considered important for the activity of PcCel45A.

3.2 Specific objectives

❖ Cloning of the PcCel45A gene from complementary DNA (cDNA) of the fungus *Phanerochaete chrysosporium*;

❖ Transformation and heterologous expression of the enzyme in *Aspergillus nidulans* (strain A773);

❖ Establishment of a protocol for purifying the enzyme from the crude extract of the fungal secretion;

❖ Biochemical characterization of the enzyme, including determination of optimum activity conditions, substrate specificity and the cleavage product released;

❖ Structural characterization through X-ray diffraction studies;

❖ Site-directed mutation in four residues of the active site, considered potentially important for the activity of the enzyme;

❖ Production of mutants D121A, D92A, Y25A and W161A;

❖ Biochemical characterization of the mutants, including specific activity tests and released products.

Chapter 4

Material and methods

"Every truth goes through three stages. At first, it is ridiculed. Then it is violently rejected. Thirdly, it is accepted as self-evident."

Arthur Schopenhauer

4 Material and methods

4.1 *Phanerochaete chrysosporium* cDNA library

For cDNA extraction, the fungus *P. Chrysosporium* was grown in the presence of 2% cellulose in Agar medium for two days at pH 4.0. Its total RNA was extracted and the cDNA was then synthesized using the Reverse Transcriptase enzyme (SuperScript® III RT, Invitrogen, CA, USA). The specific cDNA was amplified by PCR with Taq DNA polymerase according to the protocol specified by the manufacturer.

4.2 Endoglucanase V gene from *Phanerochaete chrysosporium*

The genome of the fungus *Phanerochaete chrysosporium* is deposited in the NCBI (*National Center for Biotechnology Information*) and, from it, we selected the gene with the access code (GenBank) BAG68300.1, which encoded an endoglucanase from the GH45 family. Its sequence is shown in Figure 12.

```
5' GAAACATCTCACACCGCACGAAAAGTACTATGGCGAAGCTGTCGATGTTCTT
GGGCTTCGTAGCTGTTGCGACGCTCGCGAGCGCGCTGACCGTCTCCGAGAAGC
GTGCGACGGGCGGGTACGTCCAGCAGGCTACTGGCCAGGCATCCTTCACGATG
TATTCGGGCTGCGGCTCTCCTGCTTGCGGCAAGGCTGCGTCAGGCTTCACTGCT
GCGATCAACCAGCTCGCGTTCGGCTCTGCGCCAGGCCTCGGTGCAGGCGACGC
ATGCGGACGCTGCTTCGCACTGACAGGAAACCACGATCCGTACTCTCCCAACTA
TACTGGCCCGTTCGGGCAGACGATCGTTGTAAAGGTCACGGACCTATGCCCCGT
GCAAGGCAACCAGGAGTTCTGCGGACAGACGACGAGCAACCCGACGAACCAG
CACGGTATGCCGTTCCACTTCGATATCTGCGAGGACACCGGCGGATCTGCGAAG
TTCTTCCCGTCGGGACATGGCGCGCTCACAGGCACTTTCACCGAGGTGTCGTGC
TCGCAGTGGTCGGGCTCCGACGGCGGCCAGCTCTGGAACGGTGCATGTCTCTC
TGGCGAGACTGCGCCCAACTGGCCATCGACGGCATGCGGGAACAAGGGGACT
GCCCCTTCGTAAGTTTACACTGCCCTCCATGTAGCTACGTACGATCCCTTGGATTT
CAAACAAAACTGCGTCTTTTGCAC 3'
```

Figure 12 - Sequence of the PcCel45A gene from *P. chrysosporium,* containing 717 base pairs.

Source: Prepared by the author.

4.3 Gene amplification

Forward and *reverse* oligonucleotides were used to amplify the gene of interest, the sequence of which can be seen in Table 2.

Table 2 - Sequences of the *forward* (FW) and *reverse* (RV) oligonucleotides for the amplification of the PcCel45A enzyme gene.

Name	Sequence
P. chrysosporium_FW	5' - CAGGGCGCCATGCTGACCGTCTCCGAGA - 3'
P. chrysosporium_RV	5' - GACCCGACGCGGTTACGAAGGGGCAGTCCCC - 3'

Source: Prepared by the author.

In order to correctly synthesize the oligonucleotides, we used programs that analyze amino acid sequences such as SignalP 3.0[78] and TargetP.[79] We also evaluated other characteristics of the proteins, such as the existence of rare codons[80] and glycosylations.[81]

The sequences were constructed so that the 10 nucleotides in the 5' region of the *forward* oligonucleotide and the 12 nucleotides in the 5' region of the *reverse* oligonucleotide did not have adenine in the complementary strand, so that we could use the LIC (*Ligation Independent* Cloning) cloning method, as can be seen in Table 2.

The gene amplification reaction was carried out using the enzyme *Phusion DNA Polymerase* (Finnzymes). The result of the gene amplification was verified by electrophoresis in 1% agarose gel.

4.4 Cloning using the LIC (*Ligation Independent Cloning*) system

The LIC (*Ligation Independent Cloning*) system does not require the use of restriction enzymes, digestion or even ligation reactions.[82] It is based on the 3'- 5' exonuclease activity of the T4 DNA Polymerase enzyme, which creates single strands of 10 to 15 nucleotides at the 5' end of the vector. At the same time, PCR products of the desired genes are generated, also single stranded at the 5' end, complementary to the vector. For this to happen, appropriate extensions are added to the oligonucleotides so that the vector has no thymine and the gene has no adenine in the last 10 nucleotides. Thus, both genes and vector are treated with T4 DNA Polymerase in the presence of only dATP for the gene and dTTP for the vector. Its 3'- 5' exonuclease activity will take care of removing base after base in one of the strands until it stops at a thymine (vector) and an adenine (gene), as illustrated in figure 13.

Because the vector and gene sequences are complementary, non-covalent ligation is carried out in the absence of the ligase enzyme, by incubating the DNA fragments of the gene and the vector. Ligases from the bacteria (host cell) will then covalently join the strands. This method is more efficient because it has fewer false positives, since the bacteria will only incorporate circular DNA, i.e. DNA containing the insert (gene). It is also faster and more economical, allowing many genes to be cloned at the same time.

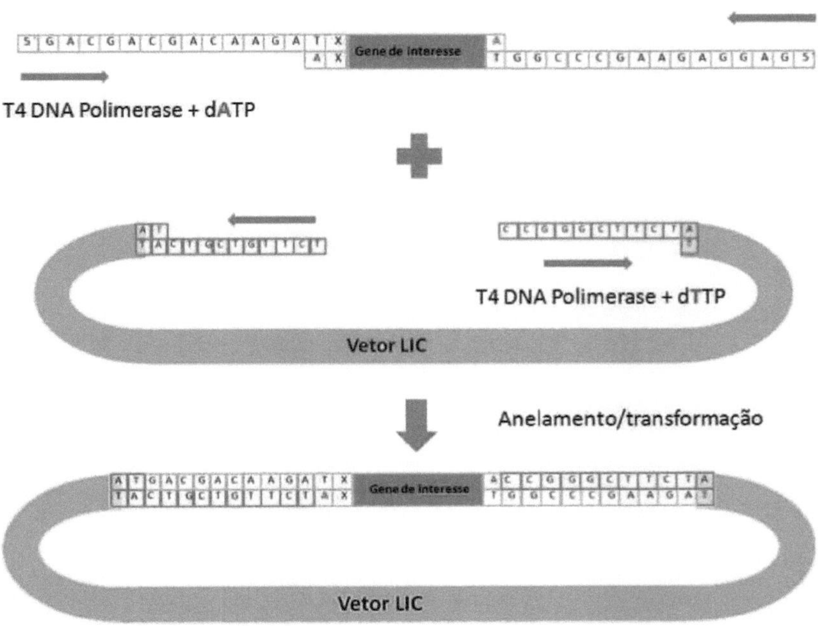

Figure 13 - Schematic representing the LIC (*Ligation Independent Cloning*) system. The production of single strands in the gene allows it to be correctly positioned in the single strand of the vector, which complement each other and are then linked by the bacteria's enzymes.

Source: Prepared by the author.

4.5 pEXPYR vector

The vector used to clone the gene of interest was pEXPYR, which is a vector modified from the *E. coli* plasmid pFE2 and was developed and kindly provided by Prof. Dr. Rolf A. Prade (University of Oklahoma, USA).[83]

This vector contains the β-lactamase gene for positive selection in *E. coli* using ampicillin. It also contains a resistance gene (Phleo) under the control of the *A. niger* β-tubulin promoter for positive selection, as well as the *A. niger* pyrG gene for recessive selection of transformants. The pyrG gene encodes oritidine-5'-phosphate decarboxylase, which is involved in uridine biosynthesis. Therefore, transformed fungi will be able to grow in the absence of uridine and uracil, while non-transformed fungi will not.

In the N-terminal region, just after the glucoamylase promoter (glaAsp), a sequence coding for a histidine tail (H6) was added, together with a cleavage site recognized by the TEV (*Tobacco Etch Virus*) protease, followed by the target protein gene. The insertion of the histidine tail facilitates the affinity purification of the protein (IMAC), while the TEV site facilitates the removal of this tail after purification.

A termination sequence (trpTC) was also added to ensure proper translation termination. With the introduction of these genetically stable proteins, whose expression is induced by maltose and secreted into the extracellular medium by the glucoamylase signal peptide, it was possible to use this vector to produce the enzyme of interest. The schematic of the vector is shown in Figure 14.

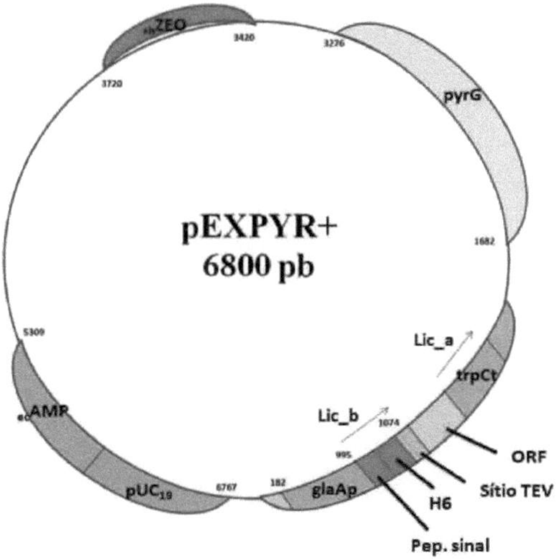

Figure 14 - Schematic representation of the pEXPYR vector. This vector has a glucoamylase promoter gene (glaAp), followed by a signal peptide, histidine tail (H6), TEV site and ORF region. It also has a trpCt termination sequence to ensure correct protein termination.

Source: Prepared by the author.

The pEXPYR vector+ was amplified using specific oligonucleotides to form the complementary strand for the LIC method. These can be seen in Table 3:

Table 3 - Sequences of the *forward* (FW) and *reverse* (RV) oligonucleotides of the pEXPYR vector.

Name	Sequence
pEXPYR/LIC _FW	5'- TGGCGCCCTGAAAATACAGGTTTTCGGTCGTTGGGATATCGTAATCGTGATGG -3'
pEXPYR/LIC_RV	5'- CCGCGTCGGGTCAAACGCTCTAGAGAACAAAAACTCATC - 3'

Source: Prepared by the author.

The vector was amplified under conditions similar to those used for gene amplification, and checked on a 1% agarose gel.

4.6 Treatment with T4 DNA Polymerase

The amplified segments were purified from the agarose gel and treated with the T4 DNA Polymerase enzyme with dATP for the gene and dTTP for the vector. The gene and vector were then incubated in the same reaction for 30 minutes at room temperature.

4.7 Transformation into *Escherichia coli* DH5α cells

The vector containing the insert was transformed into *Escherichia coli* strain DH5α, following a protocol involving heat shock.[84] For this, 10 μL of the ligation reaction was used in 50 μL of competent cell. The reaction was then left on ice for 30 minutes. For the heat shock to occur, the reaction was heated to 42 °C for 1 minute and cooled again for 2 minutes. 1 mL of LB medium was added and the reaction was incubated for 1 hour at 37 °C under stirring at 200 rpm. After this period, 200 μL of the reaction was plated onto LB-agar plates containing ampicillin as the selection antibiotic (100 μg/mL at the final concentration). The plates were placed in an oven at 37 °C *overnight.*

Positive colonies were selected by colony PCR in which the vectors with their correctly cloned inserts were in the correct *frame.*

For a more concise result, the samples were sequenced to confirm the correct cloning of the gene of interest.

4.8 Site-directed mutation

In order to carry out the site-directed mutation of the protein gene of interest, oligonucleotides were designed with the modification inserted at the 5' end of one of them and the other was designed to start in the -1 region, thus transposing the nucleotide to be modified.[85] The mutations carried out were point mutations, in the Tyr25 (Y25A), Asp92 (D92A), Asp121 (D121A) and Trp161 (W161A) residues. All residues were mutated to an alanine. The plasmid containing the gene of interest was used as a template for the PCR reaction and the enzyme *Phusion High-Fidelity DNA Polymerase* (Thermo-Scientific®). After the reaction, the product was purified and circularized using the T4 DNA-ligase enzyme. The plasmid was then used to transform *Escherichia coli* cells, as described in section 4.7.

4.9 Heterologous expression of the PcCel45A protein in *Aspergillus nidulans*

After ligating the gene into the pEXPYR vector, they were transformed into *Aspergillus nidulans* mutant A773, kindly provided by CTBE (National Bioethanol Science and Technology Laboratory, Campinas, Brazil).

To do this, the A773 mutant was inoculated in 300 mL of liquid medium containing minimal medium[86] supplemented with uridine (2.5 mg/L), uracil (2.5 mg/L) and pyridoxine (1 mg/L) at 30 °C, 200 rpm, for approximately 14 hours. The mycelium obtained was filtered and then solubilized with the enzymes "*Lysing enzymes from Trichoderma harzianum*" (Sigma L1412) and "*Lysozyme from Chicken Egg White*" (Sigma L6876) for 2 hours, under constant agitation at 200 rpm.

After the mixture became completely homogeneous and liquid, it was filtered again and the protoplasts collected. The protoplasts were then washed with "*Double Strength Protoplasting Solution*" (DSPS - 1.1 M KCl, 0.1 M citric acid, 1.0 M KOH) three times and resuspended with

STC50 solution (1.2 M Sorbitol, 50 mM $CaCl_2$, 50 mM Tris-HCl pH 7.5).

To confirm that the protoplasts had been produced, 10 μL of the solution were analyzed under a magnifying glass. Once the protoplasts were ready, 25 μL of DNA at a concentration of 1.0 μg/μL were mixed with 150 μL of protoplasts for 15 minutes. Next, the 50% PEG 4000 solution was added to the DNA and protoplasts. The mixture was then homogenized and incubated for 15 minutes. After incubation, the reaction volume was topped up to 10 mL with STC50 solution. To recover the transformed protoplasts, the mixture was plated on plates of minimal medium (absent of uracil, uridine and 5-fluoroorotic acid) and 1.2 M sorbitol, incubated at 37 °C for 4 days.

4.10 Clone selection and expression testing

When the transformed A773 clones had grown on the plate, they were selected and catalogued. The clones were duly numbered and replicated on another plate with minimal medium without sorbitol.

To find the clones "over-expressing" the protein of interest, an activity test was carried out on agar plates. To do this, plates were made with 15 g/L agar, 0.5% CMC and 0.1% AZCL-xylan, where the fungus was spiked in such a way as to allow moderate growth. To check for activity (presence of halos), the plate was filled with congo red solution (1.0 mg/mL congo red, 5 mM NaOH) and incubated for 20 minutes. A solution of 1.0 M NaCl and 5.0 mM NaOH was used for decolorization.

After checking the halos, the clones with the highest activity were transferred to 10 mL of minimal medium with maltose, which is the source of induction of the glucoamylase promoter in the pEXPYR vector construct. After 72 hours of cultivation at 37 °C, the cultures were filtered, concentrated 10 times and observed on a 15% SDS PAGE gel.

Finally, when the clone with the highest expression was found, it was inoculated into 6.0 liters of medium for the production of the enzyme of interest and subsequent purification.

4.11 Production of the enzyme Endoglucanase V by *P. Chrysosporium*

After carrying out the expression test, the positive super-expressing clone was replated on plates containing minimal medium and after incubating for 3 days at 37 °C, the spores were resuspended in saline solution (0.5% NaCl) and inoculated into liquid medium at a ratio of $2x10^6$ spores/mL. In this process, 6 liters of culture were produced, divided into 12 *erlenmeyer flasks*, and the culture was kept stationary in an oven at 37 °C for 60 hours.

4.12 Purification

4.12.1 Concentration of PcCel45A protein by Tangential Flow Filtration

The crude extract from the culture of the fungus *Aspergillus nidulans* was concentrated using *Hollow-*

fiber equipment. This equipment is a biological solution filtration system that uses hollow fiber membrane cartridges to concentrate volumes of up to 10 liters. A 5 kDa column (4 times smaller than the protein) was used so that the *P. chrysosporium* protein would not penetrate the membrane and would be concentrated in the reservoir, as seen in figure 15. To insert the culture into the cartridge, it was first filtered through a *miracloth*® membrane. At the end of the procedure, 2 liters of 20 mM Tris pH 8.5 buffer was added to change the pH of the protein, in order to continue the purification using the *Q-sepharose* ion-exchange column.

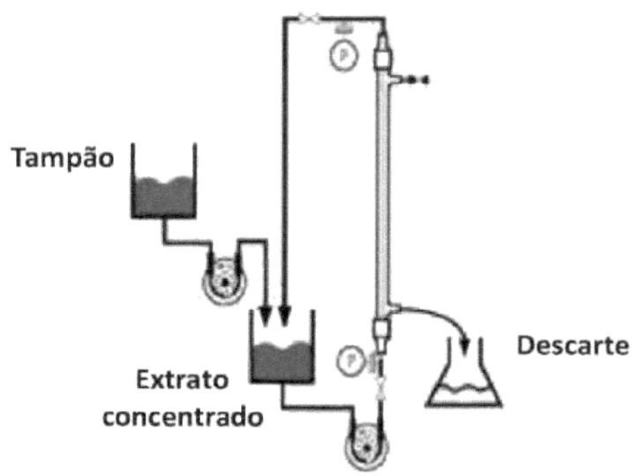

Figure 15 - Schematic representation of protein concentration in *hollow fiber*. This equipment allows the concentration of large volumes of crude extract from the culture of the fungus *Aspergillus nidulans* for the production of the enzyme of interest.

Source: Adapted from GE LifeScience[87]

4.12.2 Ion-exchange chromatography

To carry out the ion exchange chromatography, the isoelectric point of the protein of interest was researched using *ProtParam*, which is a computational tool that allows the calculation of the physical and chemical parameters of proteins through their sequence.[88]

Thus, we found that the isoelectric point of the protein of interest is 5.7, and for the chromatography, the pH used was 8.5. This gives the protein a negative electrical charge, which facilitates its adsorption to the resin (positive electrical charge) and allows the elution of proteins with the same charge.

Initially, the FF 16/60 *Q-sepharose* column was packed with the resin and washed with water. It was then equilibrated with 20 mM Tris pH 8.5 buffer, the same buffer in which the protein was stored. A volume of 100 mL of the sample was injected and elution took place with a gradual increase in the same buffer, plus 1.0 M NaCl, with the aim of altering the conditions of ionic strength, neutralizing the interactions between the protein and the resin.

The extracellular extract provided by *A. nidulans* produces few contaminants, which allowed us to finish the purification with this single step, reducing the chances of losing proteins in subsequent steps.

The eluted aliquots were subjected to 15% polyacrylamide gel electrophoresis and stained with *Coomassie Brilliant Blue* R-250 to confirm the purity of the samples.

4.13 Enzymatic tests

Initially, the enzymatic activity of PcCel45A was measured colorimetrically using CMC as a substrate (as it is the most commonly used substrate for endoglucanases) and dinitrosalicylic acid (DNS) as the chemical agent revealing the amount of reducing sugars released as a result of the enzymatic activity.[89]

The DNS method is based on the oxidation of the carbonyl group and consequent reduction of the yellow-colored 3,5-dinitrosalicylic acid to the product 3-amino-5-nitrosalicylic acid, which is reddish-orange under alkaline conditions. The result can be measured by reading it at 540 nm on a spectrophotometer.

The total volume of the reaction was 100 μL, with 50 μL of substrate (1% w/v), 10 μL of enzyme at 0.5 mg/mL and 40 μL of 50 mM buffer. Initially, we used the universal buffer: acetate/borate/phosphate pH 2.0 - 10. The reaction was maintained for 15 minutes at a constant temperature of 50 °C using a thermal cycler (BioRad).

All tests were carried out in triplicate and as a negative control (blank) we replaced 10 μL of the enzyme with 10 μL of buffer. To stop the reaction, 100 μL of DNS was added to the solution and it was immediately heated to 95 °C. The reactions were then read at 540 nm.

At the pH where the enzyme showed the highest activity, the same procedure was carried out, varying the temperature between 30 and 80 °C in a thermocycler with a temperature gradient.

Different substrates were used to test the enzyme's specificity. PcCel45A (0.1 $mg.mL^{-1}$) was incubated with 1% (w/v) solutions of lichenan from Icelandic moss, β-glucan from Barley, arabinan from beet, galactomannan, avicel PH 101 (microcrystalline cellulose), carboxymethylcellulose (CMC), tamarind xyloglucan (Megazyme), 1,4-β-d-galactomannan, laminarin and a 0.7 cm disk of Whatman N° 1 filter paper (Whatman), in a final volume of 100 μL containing 50 mM sodium citrate pH 4.0 for 15 minutes at 50 °C.

To calculate the reducing sugars released, a standard curve was constructed with different concentrations of glucose, from which it was possible to estimate the amount of sugars released by the hydrolysis of the enzyme. The calculation was made per unit of enzyme activity (U), which is the amount of enzyme that releases 1 µmol of reducing sugar per minute. The specific activity was determined as the number of U units per milligram of enzyme.

4.14 Analysis of the reaction product by TLC (*Thin Layer Chromatography)*

Thin Layer Chromatography (TLC) is a simple, fast and inexpensive method for analyzing the different compounds in a mixture. It consists of using a glass, metal or plastic container coated with a solid adsorbent plate (usually silica or aluminum).

A small amount of the mixture to be analyzed is then pipetted onto the underside of the *spot* plate, which is placed inside this container, which contains a solvent, so that only a small part of the plate is in contact with the liquid. This liquid, or eluent, is the mobile phase that slowly moves the mixture by capillary action.

As the solvent will move past the point where the mixture was applied, an equilibrium will be established for each component of the mixture between the molecules that will be adsorbed on the solid phase and those that will be in solution. In principle, the components will differ in the solubility and strength of their adsorption to the adsorbent and some will be charged higher up the plate than others. When the solvent reaches the top of the plate, the plate is removed from the solution and the drying process begins.

The hydrolysis product of a reaction containing PASC (0.5%), 3 µM of enzyme and sodium citrate buffer pH 4 (50 mM) was incubated at 50 °C for 24 hours under constant stirring at 1000 rpm. The reaction system (200 µL) was centrifuged at 13,000 rpm and the supernatant was used for analysis.

20 µL of the supernatant was applied to a plate pre-coated with silica gel 60 (Sigma). The run was started using a solution of butanol, acetic acid and water (2:1:1) as the mobile phase. The sugars were revealed using a solution of sulfuric acid:ethanol (5%) and heating to 95 °C.

4.15 Analysis of hydrolysis reaction products by HPAE-PAD

High-performance Anion Exchange Chromatography (HPAE) is used to separate anionic analytes, which are anions in their common form, or analytes that can be ionized at high pH values, such as carbohydrates at pH greater than 12.

This chromatography technique was used in conjunction with electrochemical detection by pulsed amperometry (PAD). In this method, several potentials are applied to an electrode over a specific period of time. The variations in potential are detected as a waveform, which results in the oxidation and reduction of the conditions on the electrode surface, resulting in the oxidation of the analytes adhered to the electrode surface.

In the case of carbohydrates, their hydroxyl group is oxidized on the electrode surface and the resulting current is measured. The oxidation of the carbohydrate takes place at a specific potential and results in the loss of a proton, which causes a current flow that can be measured, guaranteeing sensitive and selective detection. After oxidation, a reduction and reoxidation step is carried out to remove the bound analyte and renew the electrode surface.[90]

Therefore, the hydrolysis products were analyzed by HPAE-PAD for a more sensitive analysis of the activity of the native enzyme and its mutants. The column used was a CarboPac PAI anion exchange column (Dionex ICS - 5000, Termofischer). The mobile phase of the chromatography was a solution of 100 mM sodium hydroxide at a flow rate of 1 $mL.min^{-1}$. A solution of glucose, cellobiose, cellotriose, cellotetraose, cellopentaose and celohexaose was used as a standard.

4.16 Crystallization tests

In general, protein crystallization is a complex process and the least understood in macromolecule crystallography, followed by X-ray diffraction.[91] This is due to the fact that proteins are large, flexible macromolecules, sometimes composed of multiple subunits, chemically and physically unstable.[92] In addition, each protein is unique in its amino acid composition, resulting in a unique tertiary structure with different surface characteristics.

In general, the higher the purity of the solution containing the protein, the greater the chance of crystallization. The theory of the crystallization process involves three steps, described below and represented in figure 16:

1. The protein must be dissolved in a suitable solvent (usually a capping system), which can be precipitated by a salt or organic compound;

2. The solution is brought to supersaturation. In this step, small aggregates called nuclei of crystalline growth are formed. This process, called nucleation, represents the most difficult phase, as the molecules need to move from a disordered state to an ordered one;[93]

3. Once the crystalline nucleus is formed, crystal growth begins to occur through the diffusion of particles to the surface of the nucleus.[94]

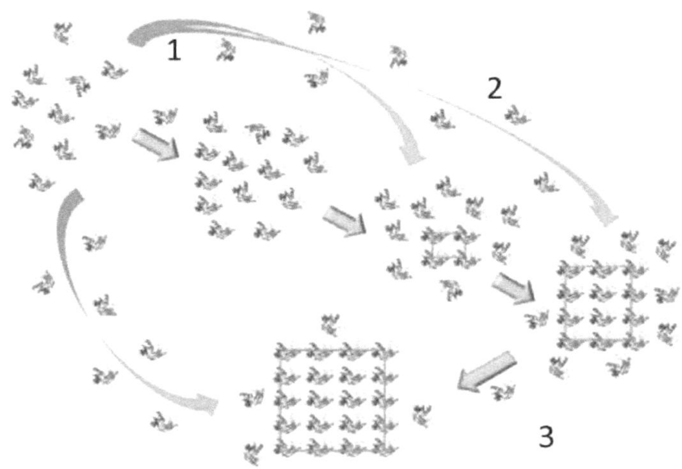

Figure 16 - Model of the formation of the crystalline nucleus. The increase in supersaturation promotes the association of molecules which begin to organize and reorient themselves, forming geometrically arranged interactions. These interactions tend to order and stabilize the aggregate, which increases in size, giving rise to the nucleus, which will develop into a crystal.

Source: Adapted from KRAUSS et al. [94]

For our crystallization tests, we initially used the sitting drop steam diffusion method, using the *Honeybee* crystallization robot and 96-well plates. In this method, a drop containing the macromolecule to be crystallized is balanced against a reservoir containing the crystallization solution at a higher concentration than in the drop. The difference in concentration between the drop and the reservoir directs the system, due to the difference in chemical potential, towards a concentration equilibrium between the drop and the reservoir solution (usually made up of cap, salt and precipitant). This happens by diffusion through the vapor phase from the less concentrated solution to the more concentrated one.

For an initial test, we *screened* different conditions of commercially available crystallization solutions, including *PEG I* and *PEGII*, *SALTRX* and *Crystal Screen*. The conditions that proved promising were then refined manually using the hanging drop method in 24-well plates. This method is also based on steam diffusion, differing from the seated drop only in the position of the drop. We used 1 μl of protein and 1 μl of crystallization solution on a siliconized coverslip. The coverslip was inverted and sealed over a reservoir containing 1000 μl of crystallization solution (Figure 17).

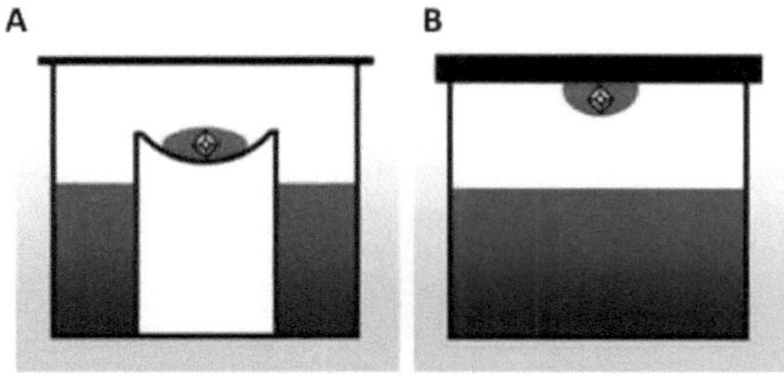

Figure 17 - Protein crystallization methods. In A: *sitting drop* method. The protein solution (in pink) is placed on a pedestal together with the crystallization solution on the reservoir containing the crystallization solution. In B: *hanging drop* method. The protein solution is placed on a coverslip together with the crystallization solution and this is inverted over the reservoir containing the crystallization solution.

Source: COLLABORATIVE Crystallization Center (C3)[95]

The initial protein concentration used in these crystallization tests was 14.5 mg.mL^{-1}.

4.17 Collection and processing of crystallographic data

The structural crystallography of macromolecules is based on the diffraction of X-rays by the electrons of the atoms that make up the molecules present in the crystal. As the similar structural motifs that form the unit cell are repeated throughout the volume of the crystal on a periodic basis, it is possible to reconstruct the data obtained to form a 3D structure.

The geometry (size and shape) of the unit cell and the wavelength of the X-rays (from 0.8 to 1.6 Â), which is in the same range as the interatomic distances of the molecules in the crystal, allow for increased scattering of the radiation in certain regions (constructive interference) and extinction in others (destructive interference), giving rise to the diffraction pattern observed in X-ray diffraction experiments on a macromolecule crystal.

For the apo protein, prior to the X-ray diffraction experiments, the crystals were cryoprotected by adding 10% (v/v) ethylene glycol to the crystallization solution and collected in a stream of nitrogen vapor at cryogenic temperature (100 K). The X-ray diffraction data were collected at the *National Synchrotron Light Source* (NSLS, USA) during the course of RapiData (2013) by our group's PhD student Andre Godoy. The experiments were carried out on the X4C beamline, equipped with the MAR 165 CCD detector, at a wavelength of 0.98 Â.

In order to obtain as many measured reflections as possible, it is necessary for different pianos of the crystal lattice to enter into diffraction conditions. To do this, we use the rotation method, in which the wavelength (λ) is fixed and the crystal is rotated around a single axis, so that at each moment, new planes enter diffraction conditions.

After obtaining the first reflections, the images were indexed. In this step it is possible to determine the parameters independent of the phase information, such as the orientation and size of the unit cell, the point group, which designates the symmetry of the crystal, and to estimate the mosaicity. By carrying out this step at the beginning of the collection, it is possible to plan a strategy that maximizes the quality of the results, minimizing the exposure time of the crystal in the X-ray beam.

Once the images had been taken, the data collected was integrated using the XDS program.[96] In this step, the intensity of each measured reflection is determined. Due to factors such as variations in the incident radiation, absorption of diffracted rays or destruction of the crystal, the intensities of each point are recorded on different scales. In order to group the data together, they need to be on the same scale, which is why the data were scaled. After this step, it is possible for equivalent and symmetrical reflections from different images to have their intensities added together, allowing the data to be merged.

The structure of the enzyme in complex with cellobiose was determined using the *soaking* technique, where the native protein crystals used for data collection were dipped in a saturated solution of cellobiose prior to the X-ray diffraction experiments. After 3 hours, the crystals were collected in a Bruker APEX DUO diffraction system equipped with a KAPPA goniometer and APEX II CCD detector. The exposure time was 10 seconds per *frame. The* data was integrated with SAINT and scaled via SADABS.[97] The statistics of the data collection and the determination of the spatial group were obtained with XPREP, [97] and are shown in table 4.

4.18 Structure resolution

The structural model of PcCel45A was obtained by molecular replacement,[98] using a pair-wise comparison method with available crystal structures, using the Markov algorithm.[99] The best model found for solving the structure was that of *Mytilus edulis* endoglucanase (Me_EGV, PDB 1wc2) with only 20% sequence identity with PcCel45A. This low sequence identity meant that the model initially generated needed to be adjusted. To do this, the electronic density was modified using the RESOLVE[100] program and submitted to AutoBuild.[100] The refinement aims to find a better agreement between the proposed model for the structure of the macromolecule and its real structure. This agreement lies in the equality between the calculated structure factors (F_{calc}) and the observed structure factors (F_{obs}).

The quality of the refinement process is therefore monitored by calculating the R_{factor}, which measures the discrepancy between F_{calc} and F_{obs}. To avoid an artificial minimization of the Rfactor, which can be achieved through an over-refinement process, a small percentage of the reflections are excluded from the refinement and used as a test set for validation in the calculation of a new R-factor, called

Rfree.

The final model was then built and refined with the programs Coot[101] and phenix.refine[102] and validated by the program MolProbity.[103]

Chapter 5

Results and discussion

- Obtaining the pure enzyme and its characterization

" Life only gives wings to those who are not afraid to run. "

Kleber Martins

5 Results and Discussion

5.1 Cloning

The PcCel45A gene was cloned by the LIC system using the pEXPYR vector. The amplification reaction was confirmed using 1% agarose gel electrophoresis (figure 18). The newly constructed recombinant plasmid was then transformed into *E. coli* strain DH5α for propagation and the plasmid DNA was extracted. This propagation and DNA extraction is necessary to obtain high concentrations of the plasmid (above 500 ng/µL), which is essential for the transformation process in *Aspergillus nidulans*.

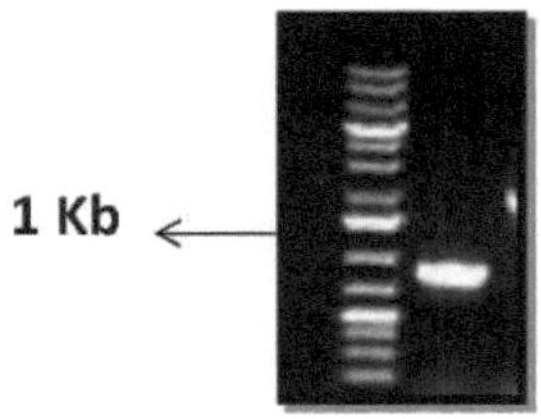

Figure 18 - 1% agarose gel showing the amplification of the PcCel45A gene with 717 bp from the *P. chyrsosporium* cDNA. The first column refers to the molecular mass marker.

Source: Prepared by the author.

The use of the ligase-independent cloning (LIC) technique as a cloning strategy proved to be very efficient and with few false positives.

5.2 Heterologous expression in *Aspergillus nidulans*

The initial expression test was carried out on the plate with Agar and CMC, in which information was obtained from the colonies that showed the most activity (largest halos on the plate), as can be seen in figure 19.

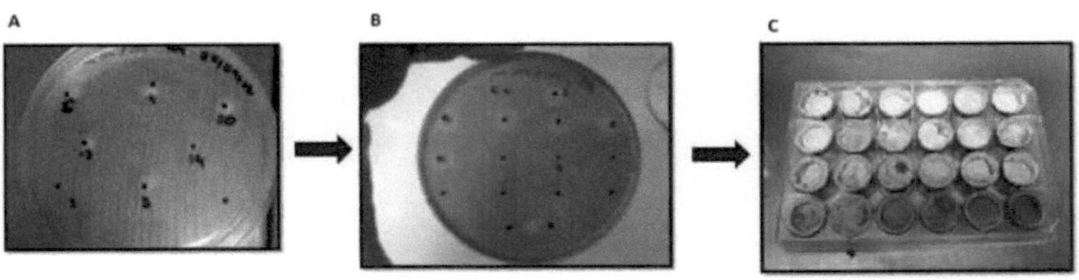

Figure 19 - Expression test carried out after transforming the PcCel45A enzyme into *A. nidulans*. A) Cataloged colonies. B) Expression test in medium containing CMC. After checking the halos (enzyme activity), these colonies were transferred to liquid medium (C). Expression was then checked by SDS-PAGE gel.

Source: Prepared by the author.

From this test, the colonies with the largest halos were transferred to 10 mL of minimal medium with maltose, incubated for 60 hours and the result was analyzed by SDS-PAGE (figure 20).[104] By observing the expression of the colonies on the gel, it was possible to choose the colony with the highest activity and the fewest contaminants, which was chosen for large-scale production of the enzyme.

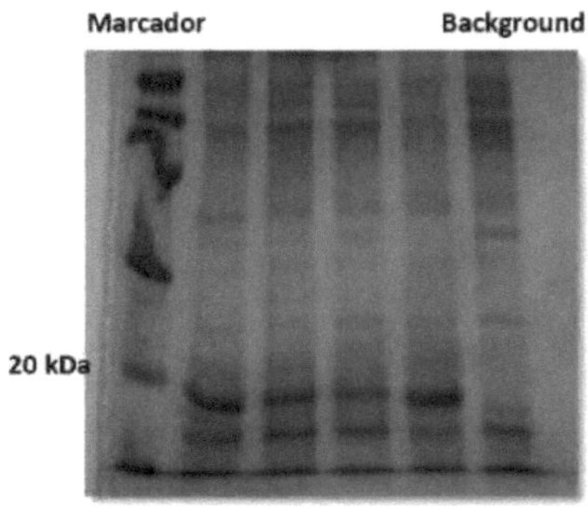

Figure 20 - Analysis of the expression of PcCel45A by four colonies previously selected by the top agar experiment. The comparison of protein production between them allowed us to obtain the super-expressor that was chosen for large-scale protein production.

Source: Prepared by the author.

5.3 Expression of endoglucanase from *P. Chrysosporium* by *A. nidulans*

The spores were inoculated from the plates of the "super-expressor" clone *of A. nidulans* into 6 liters of culture containing minimal medium, 1 mg/L pyridoxine, 3% maltose and 100 mM phosphate buffer to maintain the medium at pH 6.5. The medium was kept stationary in an oven at 37 °C for 60 hours. After incubation, the extracellular medium was separated from the fungus growing on the surface by filtering it through *a Miracloth*® membrane (Figure 21).

Figure 21 - *A. nidulans* culture after a 60-hour incubation for large-scale protein expression. The extracellular medium is separated from the fungus growing on the surface by filtration and then concentrated in a *Hollow-fiber* cartridge.

Source: Prepared by the author.

5.4 Purification

5.4.1 Protein concentration in *hollow-fiber* cartridge

After cultivation, the medium was concentrated in a *hollow-fiber* cartridge to approximately 100 mL. The cap was changed by adding 20 mM Tris pH 8.5 to the cartridge and allowing the concentration to reach 100 mL again. The result was observed on an SDS-PAGE gel, as can be seen in Figure 22. Despite the existence of some contaminants, it can be seen that the most expressed protein is PcCel45A.

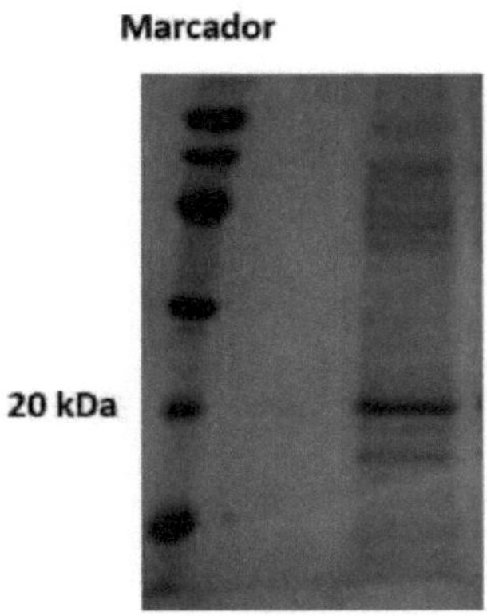

Figure 22 - SDS-Page 15% relative to the concentration of 6 liters *of hollow fiber* culture. After concentrating the medium, the most prominent band is the PcCel45A protein.

Source: Prepared by the author.

5.4.2 Protein purification by ion-exchange technique on a *Q-sepharose* column

The concentrated extract was then applied to a *Q-sepharose* column in an *Àkta Purifier* system (GE Healthcare®). After the run was complete, two 10 mL fractions of the peak were eluted. These were applied to an SDS-PAGE gel and we were able to verify their purity, as can be seen in Figure 23. To carry out the crystallographic tests, the fractions were concentrated in 10 kDa *vivaspin* (*Amersham Bioscience*) to a final concentration of 14.5 mg/mL.

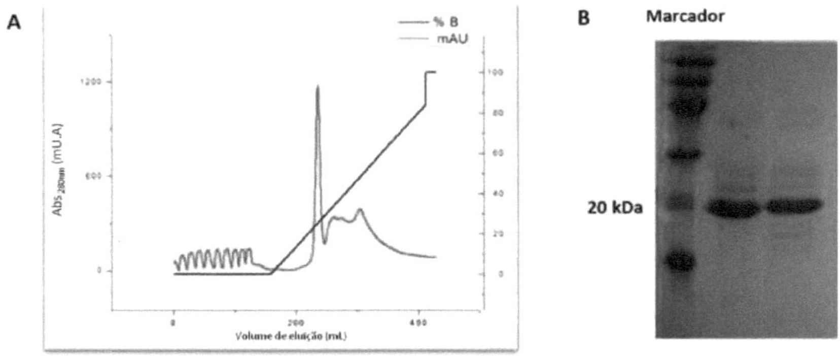

Figure 23 - Purification of the enzyme PcCel45A. A) Ion-exchange chromatogram on a 16/60 Q-sepharose column of the purification of PcCel45A. The peak in red corresponds to the elution of the protein. In the legend, %B corresponds to Sodium Citrate, pH 4.0, plus 1 M NaCl. B) SDS-PAGE 15% showing the purity of the samples after ion-exchange purification.

Source: Prepared by the author.

5.5 Biochemical characterization

After obtaining the pure protein, we began the biochemical characterization of the enzyme.

5.5.1 Optimum pH test

Initially, a pH *screening* was carried out at a temperature of 50 °C. This temperature was chosen because it is the most common optimum temperature for cellulases. Analysis of the results obtained showed that the PcCel45A enzyme has its highest activity at pH 4.0, confirming its acidophilic profile like other enzymes in the family and fungal cellulases in general. At pHs higher than 4.0, its activity decreases drastically and it is practically inactive from pH 7.0 onwards. The activity profile of PcCel45A as a function of pH, for determining the optimum pH, can be seen in Figure 24.

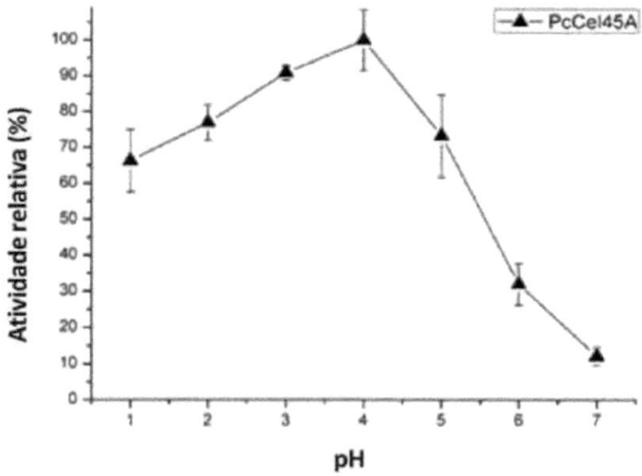

Figure 24 - Profile of the relative activity of the PcCel45A enzyme for the different pHs. Keeping the temperature constant, the pH range was varied between 1 and 10, and from pH 7 onwards the enzyme no longer showed any significant activity.

Source: Prepared by the author.

5.5.2 Temperature test

After analyzing the pH at which the protein was most active, we carried out the same test by varying the temperature. PcCel45A had its optimum activity at around 70 °C, a result close to the optimum temperatures of other enzymes from this fungus.[15] Comparing our results with enzymes from the 45 family, we noticed that the endoglucanase from *Syncephalastrum racemosum* (subfamily A)[105], and the endoglucanase from *Melanocarpus albomyces* (maEG)[(67)] also have an optimum temperature of 70 °C. Other enzymes such as *Penicillium decumbens* endoglucanase[106], *Staphylotrichum coccosporum* endoglucanase[107] and *Trichoderma reesei* endoglucanase[108] showed optimum activity at 60 °C. The results of the optimum temperature for PcCel45A can be seen in Figure 25.

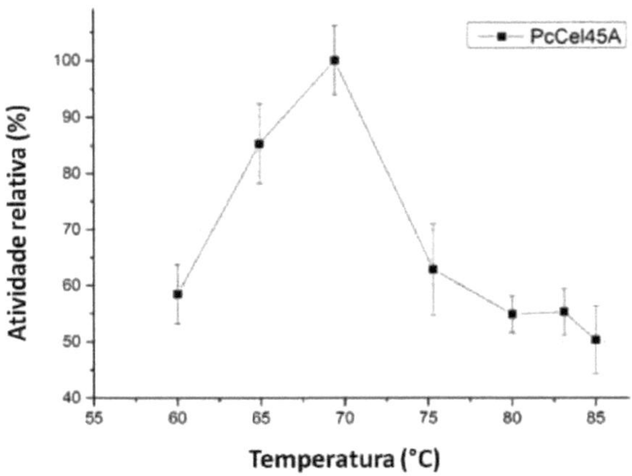

Figure 25 - Profile of the relative activity of the PcCel45A enzyme varying the temperature from 30 to 80 °C. For this test we maintained the enzyme's optimum pH (pH 4.0).

Source: Prepared by the author.

Under optimum pH and temperature conditions in Iiquenano, the enzyme showed a specific activity of approximately 1.22± 0.03 U mg^{-1}. This value, although lower than that found for most endoglucanases from other families, is close to that reported for some enzymes from the same family, such as the endoglucanase from *Neurospora crassa.*[109]

5.5.3 Panel of substrates

The enzyme had its specificity tested against different substrates, showing activity on the following: lichenan, followed by filter paper (approximately 80% of the activity relative to the maximum observed on Iiquenan), β-glucan (38%), galactomannan (11%) and CMC (10%), as can be seen in figure 26.

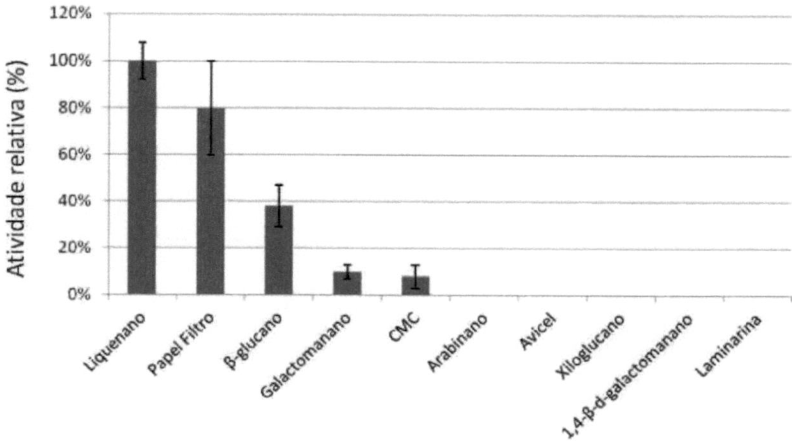

Figure 26 - Substrate specificity profile of the PcCel45A enzyme. Histogram represented in terms of relative activity after 15 minutes of reaction under optimal conditions with the different substrates. Standard deviation referring to three independent experiments.

Source: Prepared by the author.

The biochemical characterization of this enzyme had already been carried out by another group,[12] and to refer to it we will use the name PcCel45App. The aim of the tests with PcCel45A was to compare the results, because although they were the same enzyme, the expression systems in which they were produced were different, and the other group used *Pichia pastoris*.

For the characterization of PcCel45App, MCC (microcrystalline cellulose), PASC, CMC, lichenan, β-glucan from barley, glucomannan and xylan were used. The results can be seen in figure 27.

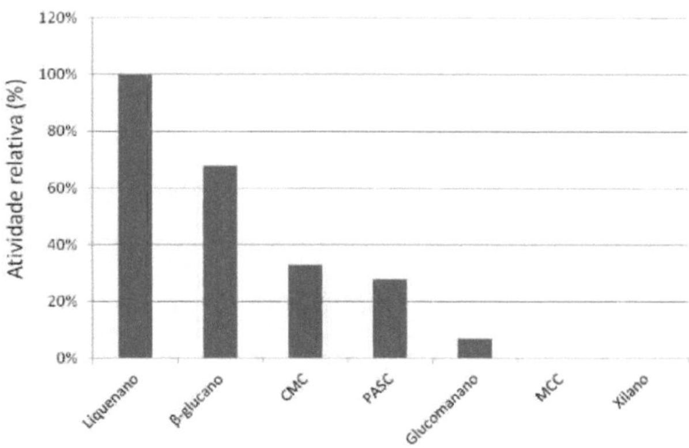

Figure 27 - Substrate specificity profile of the PcCel45pp enzyme. Histogram represented in terms of relative activity, after 1 hour of reaction, at a temperature of 30 °C.

Source: Adapted from IGARASHI.et al[12]

Comparing the results, we found that both enzymes showed greater activity for the Iiquenane substrate, which is made up of β1,3/β1,4 bonds in a 1:2 ratio. The filter paper substrate appeared in our tests as the second highest in enzymatic activity, and was not tested by the other group. Filter paper is composed of β1,4 bonds, and differs from avicel (microcrystalline cellulose) in that it has a higher degree of polymerization and lower crystallinity.

Both enzymes showed activity towards β-glucan (β1,3/β1,4 bonds in a 1:4 ratio) and CMC (β1,4 bonds). Galactomannan was used for PcCel45A, while glucomannan was used for PcCel45App. The enzyme was active on both substrates. The difference between them lies in the main skeleton, galactomannan being made up of galactose, while glucomannan is made up of glucose.

The results from the other group showed that PcCel45App has greater activity for substrates with mixed bonds (β1,3/β1,4), represented by lichenan and β-glucan, than for substrates with β1,4 bonds. The results for PcCel45A are similar, as both enzymes showed greater activity for lichenan, β-glucan, CMC and gluco/galactomannan. However, PcCel45A showed greater enzymatic activity using filter paper as a substrate compared to β-glucan, suggesting that the substrate's macrostructure and the composition of the glycosidic bonds influence enzymatic activity.

To verify the action of PcCel45A, we also tested Iaminarin, which is a polysaccharide made up of glucose monomers joined by bonds (β1-3), with branches (β1-6). Although the skeleton of this substrate is made up of (β1-3) bonds, the enzyme showed no activity, which indicates that the cleavage of the enzyme in mixed substrates occurs at the β1,4 bonds.

The classic substrate preferences of this family describe that the enzymes show the highest activity on lichenan and β-glucan, followed by a lower activity on crystalline cellulose (avicel and filter paper). In addition, they have no activity on xylan, laminarin, cellobiose or synthetic substrates.[64] Although most substrate specificity tests are linked to enzymes from subfamily A, we believe that this enzyme has a very close substrate preference, which is interesting considering that they are very different enzymes.

5.5.4 Analysis of the native enzyme product by TLC

To analyze the product of the PcCel45A reaction, we used β-glucan and PASC. For β-glucan, we were unable to observe the accumulation of celooligosaccharides. With the use of PASC, however, it was possible to observe celooligosaccharides with polymerization degrees of 3, 4 and 5, as can be seen in Figure 28.

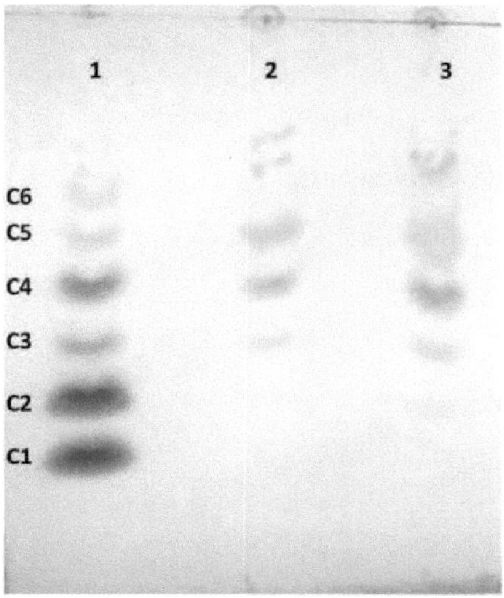

Figure 28 - Analysis of the soluble reaction products after incubation of the PcCel45A enzyme with PASC. The reaction was maintained at 50°C for 24 hours. In row 1: standard Celooligosaccharides with degree of polymerization from 1 to 6 (C1 to C6). In row 2: supernatant from the hydrolysis reaction. In row 3: reaction supernatant concentrated 10 times.

Source: Prepared by the author.

Analysis of the cleavage product using β-glucan by both TLC and HPAE-PAD showed that the enzyme does not produce oligosaccharides when there are β1,3/β1,4 bonds. This is probably because the enzyme may not recognize the repetitive units of lichenan and β- glucan, which are cellotrioses connected by β1,3 bonds. Thus, only partial hydrolysis may occur for these substrates, as has been suggested by another group.[12]

Chapter 6

Results and discussion

- Obtaining the structure of the enzyme and its uniqueness

"The most beautiful thing that man can experience is mystery. It is this fundamental emotion/that lies at the root of all science and/or all art/. "

Albert Einstein

6 Results and Discussion

6.1 Crystallization, collection and processing of crystallographic data

The PcCel45A crystals were obtained in a solution of 0.5 M ammonium sulphate; 0.1 M Hepes pH 7.5; 30% v/v (+/-) 2-methyl-2,4-pentanediol. After collection, the data was integrated into an orthorhombic point group, with the space group being P212121, with unit cell parameters equal to 45.6 Â, 58.6 Â and 63.7 Â. The crystal used for the X-ray diffraction experiment can be seen in figure 29.

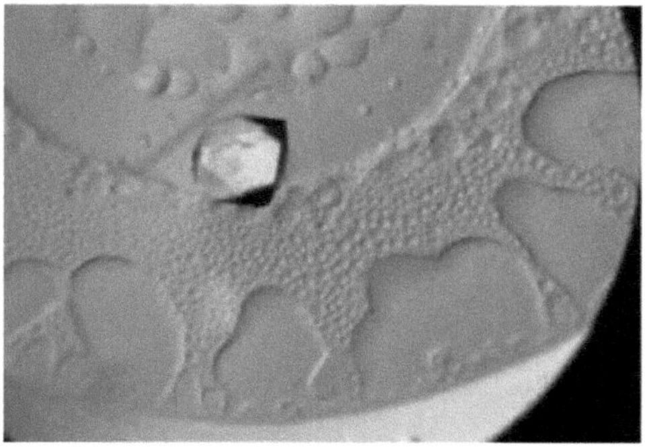

Figure 29 - Crystal of the PcCel45A protein used in the X-ray diffraction experiment.

Source: Prepared by the author.

The content of the asymmetric unit was estimated using the Matthews coefficient (Vm = 2.35 $Â^3.Da^{-1}$) which indicated the presence of only one PcCel45A molecule in the unit cell and 28% solvent. The crystallographic structure of the "apo" enzyme was refined, resulting in a final R_{work} of 15.1% and R_{free} of 17.8% at 1.4 Â resolution. The final model consists of 180 protein residues, 230 water molecules and 16 ethylene glycol molecules. After refinement, 97% of the residues were in regions allowed by the Ramachandran graph. The structure of PcCel45A has two cysteines (64 and 94) with bad angles (SG- SG), possibly due to radiation damage. The PcCel45A-cellobiose complex was refined to 1.7 Â, and is of similar quality to the "apo" enzyme. Both structures are complete from residues 8 to 187. Despite being expressed by the fungus *Aspergillus nidulans*, no post-translational modifications were found in the structure of the recombinant enzyme. The crystallographic data and refinement statistics are shown in Table 4.

Table 4 - Crystallographic data collected and refinement statistics for the PcCel45A protein in native form and in complex with cellobiose. The values in brackets refer to the last layer of resolution.

	PcCel45A	**PcCEL45A+CELOBIOSIS complex**
DATA COLLECTION		
Detector	MAR 165 CCD	APEX II CCD detector
Wave length (A)	0,97924	1,5418
Resolution range (A)	31,84 (1,47)	36,74 (1,70)
Space group	$P2_12_12_1$	P212121
Unit cell (a, b, c) (A)	45,6; 58,6; 63,7	45,3; 58,2; 62,9
Unique reflections	29354 (2568)	18567 (1694)
Multiplicity	7,4 (2,4)	5,7 (3,5)
Completeness (%)	98,22 (87,02)	99,08 (92,12)
Mean I/sigma(I)	12,50 (3,04)	8,42 (1,86)

R-merge (%)	4.3 (17.9)	37.65 (11.35)
REFINEMENT PARAMETERS		
R_{work}(%)	15,11	17,74
R_{free} (%)	17,83	20,68
Binders	16	28
Waters	230	271
Amino acid residues	180	180
RMSD (links) (A)	0,016	0,014
RMSD(angles) (°)	1,52	1,23
Ramachandran favorable (%)	97	97
Ramachandran unfavorable (%)	0	0
AVERAGE B-factors		
Protein, global (A^2)	15,60	14,00
Binder (A2)	29,30	22,40
Solvent (A2)	28,80	26,10

Source: Prepared by the author.

6.2 Structural model of PcCel45A

The general shape of the PcCel45A structure resembles an anchor, with a large slit on the surface of the enzyme, approximately 40 Â long, 13 Â deep and 12 Â wide, as can be seen in figure 30.

Similar to the other existing structures in the 45 family,[67] the structure of PcCel45A is characterized by the presence of a double-Ψ-β-barrel, made up of six β-sheets, four α-helices and nine *loops*. The structure also contains five disulfide bridges between cysteines 28 and 123, 33 and 61, 64 and 149, 94 and 103, 165 and 179, respectively.

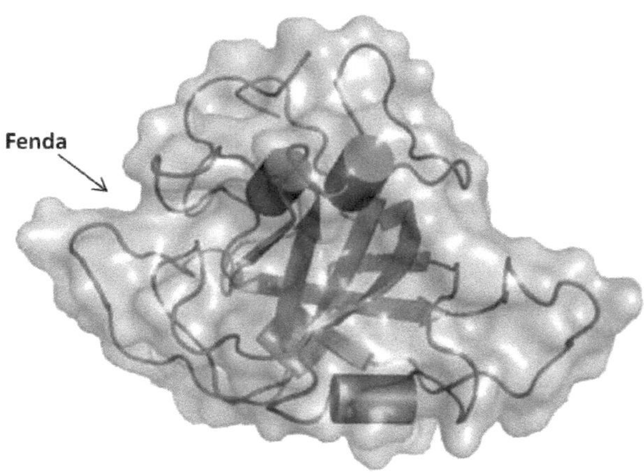

Figure 30 - General structure of PcCel45A, with emphasis on the cleft on the enzyme's surface. The general shape of PcCel45A resembles an anchor, due to the positioning of *loops* 4 and 9.

Source: Prepared by the author.

A detailed analysis of the structure shows that the β1 sheet (Thr17 - Tyr25) is connected to β2 (Thr41 - Asn45) by the second *loop,* and both are in the enzyme's active site. Between them is β5 (Phe118 - Cys123), which is parallel to the β1 sheet and antiparallel to β2, forming the groove. A disulfide bridge between Cys28 and Cys123 stabilizes the connection between the β1 and β5 sheets. The β3 (Gly62 - Gly69) and β6 (Thr143 - Glu146) sheets are located in the groove.

on the opposite side of the barrii and its residues do not interact directly with the active site (Figure 31).

Starting after *loop* 4, the N-terminal region of the β4 sheet (called β4' and comprising residues Ile86 to Val90), is anti-parallel to β3. After reaching the end of the barrii, the β4 sheet forms a kink at the Thr91 position at an angle of approximately 90° and continues as the β4"(Asp92- Cys94) sheet. The disulfide bridge between the Cys94 and Cys103 Iateral chains of helix 2 stabilizes this unusual kink in the enzyme's main chain. The β4" sheet not only forms part of the surface of the active site on the opposite side of the β1 sheet, but also contains residues that may be important for the enzyme's activity, such as Asp92 (Figure 31A).

Together with loops 2 and 3, *loop* 9 (Trp152 - Ser187) seals the other side of the groove, in the vicinity of the β4" leaf. This is the largest *loop* in the PcCel45A structure, which forms two clamps, the first (Trp152 - Cys165) being strangled by a disulfide bridge between Cys165 and Cys179 and where the Trp161 amino acid is located, the only surface tryptophan in the structure. The second clamp (Cys165 - Ser187) is part of the C-terminal region of the enzyme.

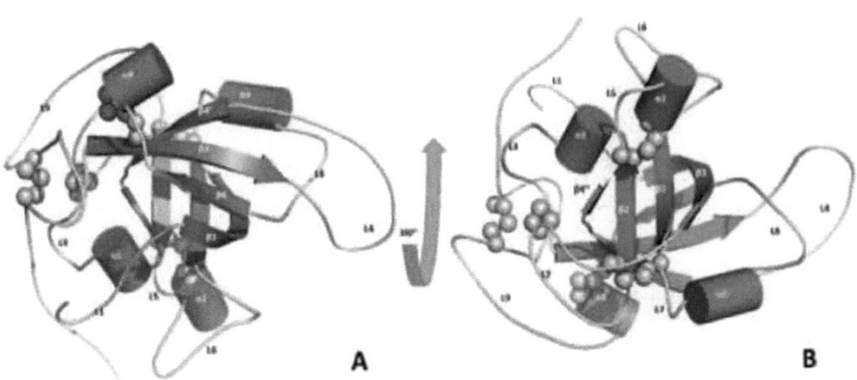

Figure 31 - General shape of the PcCel45A structure. The 6 leaves of the β-barrel structure are shown in green, followed by the 4 α-helices in orange. The *loops* are in yellow. The purple spheres symbolize the disulfide bridges.

Source: Prepared by the author.

6.3 Structure of PcCel45A in complex with cellobiose

The PcCel45A-cellobiose complex revealed a single cellobiose in the cleft of the active site. Like other GH45s, the active site is subdivided into six subsites in relation to the cleavage of the carbohydrate: +1 and +2 are at the reducing end of the carbohydrate and - 1, -2, -3 and -4 are at the non-reducing end of the carbohydrate.[56] The cellobiose molecule of the PcCel45A-cellobiose complex is in the +1 and +2 subsites as can be seen in Figure 32A.

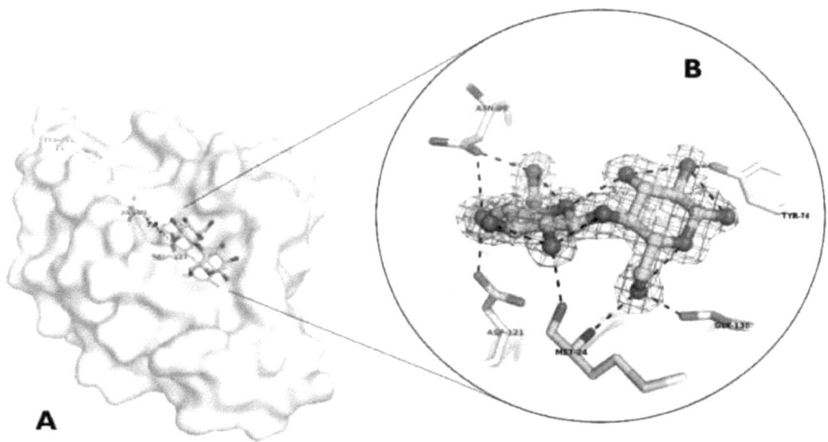

Figure 32 - PcCel45A active site. A) Emphasis on the groove on the surface of PcCel45A and the positioning of cellobiose in the +1 and +2 subsites. In the background are the amino acids Asp92 and Asp121, at a distance of 8 Â from each other. We also see the conserved tryptophan 161 residue in the C-terminal region of the enzyme. B) Emphasis on the interaction of cellobiose with the residues Asn99, Asp121, Met24, Gly138 and Tyr74.

Source: Prepared by the author.

The residues Asp121, Met24, Gly138 and Tyr74 interact directly with cellobiose through hydrogen bonds, as shown in Figure 322B. In the +2 position, the amide nitrogen of Met24 and the carbonyl oxygen of Gly138 interact with the O6' of cellobiose,

possibly by stabilizing the ligand in the active site. The O2' of cellobiose forms a hydrogen bond with the hydroxyl group of the phenyl aromatic ring of Tyr74 (2.9 Â). In the +1 position, the O2 hydroxyl of cellobiose forms a hydrogen bond with the carbonyl oxygen of Met24.

The comparison between the apo-PcCel45A structure and its complex with cellobiose revealed that the ligand binding induces a small conformational change in the β1 and β5 sheet, where the Met24 and Asp121 residues are respectively located. Even so, no significant difference was noted between the apo enzyme and its complex (RMSD of 0.12 Â for Cα).

6.4 Comparison between PcCel45A and other GH45s

As already mentioned, GH family 45 doesn't have much structural information, with most of the available structures belonging to subfamily A (Hi_EGV, maEG and HG_EGV).

Although the secondary structure elements (double-Ψ-β-barrel) are conserved in all GH45s, when we compare PcCel45A with the proteins of subfamily A, we see that it has significant differences. The RMSD between the α carbons of PcCel45A and Hi_EGV is 12 Â and with maEG 15 Â and HG_EGV it is 16 Â. They also differ in shape, as the structures of Hi_EGV and maEG (which are the best studied) have the shape of a sphere with a slit surrounded by *loops,* while PcCel45A has the shape of an anchor, where the slit crosses the surface of the structure. The comparison between the Hi_EGV and PcCel45A enzymes can be seen in figure 33.

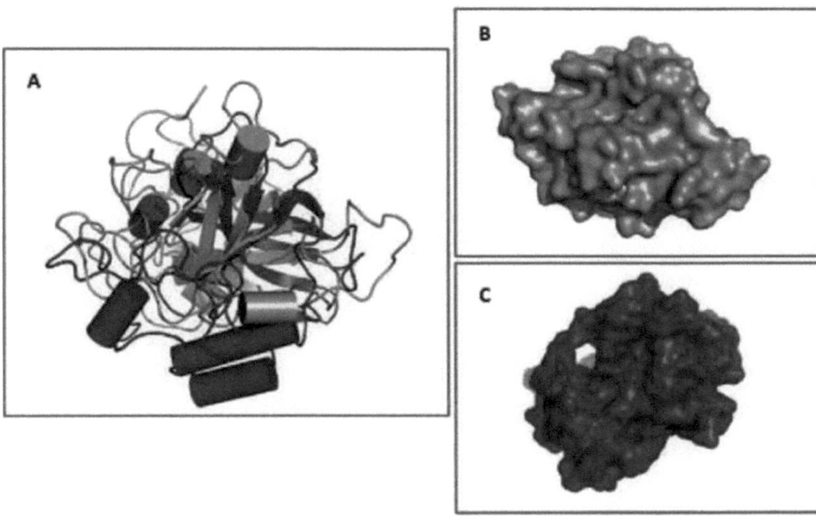

Figure 33 - Comparison between PcCel45A and the Hi_EGV enzyme. In A, the overall structure of PcCel45A (in olive green) superimposed on the structure of Hi_EGV (in blue). In B, surface of PcCel45A showing the active site groove. In C, surface of Hi_EGV, showing the active site, with *loops* surrounding the cleft.

Source: Prepared by the author.

In subfamily A enzymes, the long loops surround and hold the substrate in the active site, while the attachment of the substrate to the active site of PcCel45A seems to occur through aromatic residues on the surface, such as Tyr25, Tyr74 and Trp161.

Comparison PcCel45A with the only subfamily B structure available (ME_EGV) shows that *there is 20%* sequence identity (RMSD: 3.1 Â for Cα). PcCel45A has a more elongated cleft, because *loop* 4 is positioned externally, whereas in ME_EGV this *loop* is internal, giving this structure a more globular shape, similar to those of subfamily A. The comparison between the structures can be seen in Figure 34.

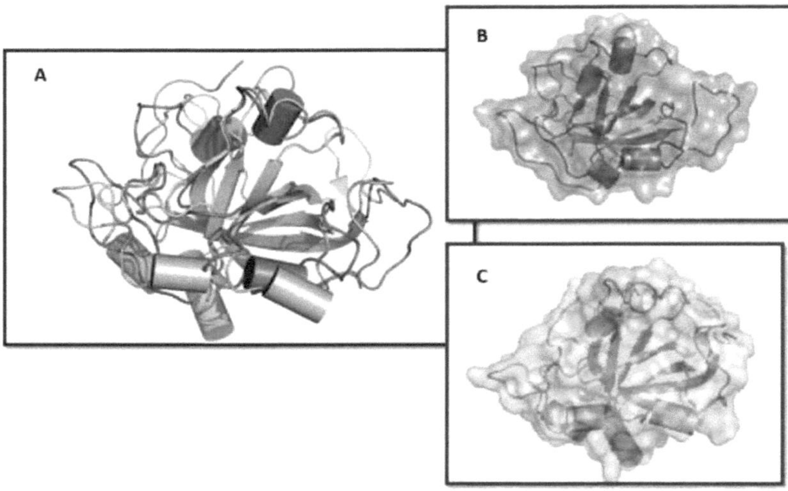

Figure 34 - Comparison of the ME_EGV enzyme with PcCel45A. A) Superimposition of the two structures. B) General shape of the PcCel45A surface, resembling an anchor. C) General shape of the ME_EGV surface, showing a less elongated and more globular cleft.

Source: Prepared by the author.

In addition to the unique morphology of the PcCel45A structure, another important difference found between this enzyme and the other representatives of the family is the fact that Asp10, known as the proton acceptor in the inversion reaction, does not exist in PcCel45A (Figure 3535).

Instead, PcCel45A has another aspartate, Asp92, which is located in the cleft (β4" sheet) and is 8 Â away from Asp121. However, compared to the catalytic aspartates of Hi_EGV, for example, this PcCel45A residue is positioned linearly to Asp121, as seen in Figure 3535. This position would make it difficult for the substrate to be cleaved by the inversion reaction.

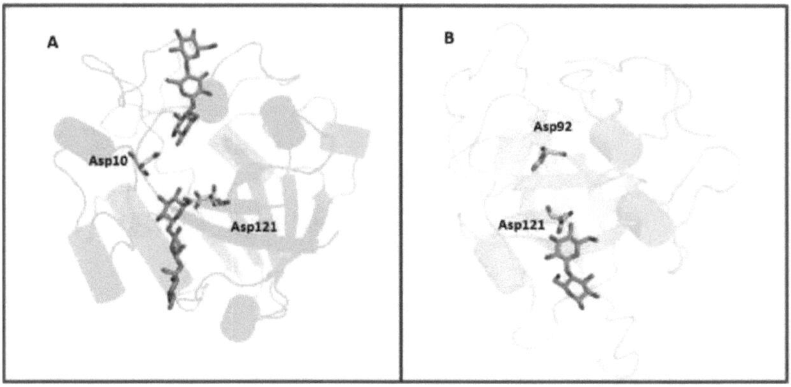

Figure 35 - Comparison between the structures of Hi_EGV and PcCel45A, highlighting the aspartates. A) Emphasis on the catalytic aspartates (yellow) present in the active site of Hi_EGV in complex with celohexaose (green). B) Emphasis on the aspartates of the PcCel45A site (in yellow) in complex with cellobiose (green).

Source: Prepared by the author.

Another residue that is conserved in the Hi_EGV and maEG structures and is also found in PcCel45A is Tyr25. This residue is located at the end of the β1 sheet and, as in the other structures, forms the floor of the catalytic center, as shown in Figure 36.

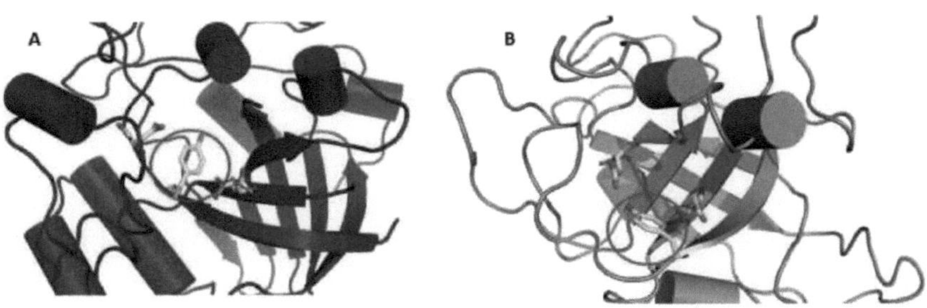

Figure 36 - Active site of Hi_EGV and PcCel45A. A) Amino acids Asp121, Asp10 and Tyr8 in yellow, highlighting Tyr25 which makes up the floor of the catalytic center in Hi_EGV. B) Amino acids Asp121, Asp92 and Tyr25 in the structure of PcCel45A.

Source: Prepared by the author.

Analyzing the other representatives of GH45 in complex with the substrate and PcCel45A in complex with cellobiose, we see that the positioning of the ligand in the cleft is similar in all the enzymes, with Asp121 bound to the +1 subsite of the substrate, which suggests that cleavage by PcCel45A also occurs between the -1 and +1 subsites and that Asp121 is in fact the proton donor for the reaction.

6.5 Structural comparison of PcCel45A and β-expansins

It is already known from phylogenetic studies that the enzymes of family 45 have a distant link with plant expansins.[12] In a comparison between PcCel45A and domain one of the expansins, numerous similarities were found. The structural comparison with an expansin known as β-expansin from the group 1 subfamily of allergenic grass pollen (EXPB1) showed an RMSD of 0.87 Å (for 138 atoms) and a sequence identity of 15%. The superposition of the structures can be seen in Figure 37.

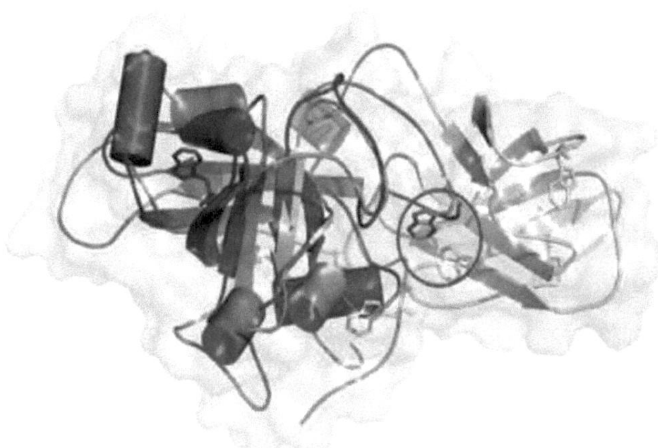

Figure 37 - Superimposition of the structure of PcCel45A (in olive green) with domain 1 of the EXPB1 expansin (red), with emphasis on the tryptophans. Circled in red is Trp161 of PcCel45A superimposed with domain 2 of the expansins.

Source: Prepared by the author.

As we can see in figure 37, the structure of domain 1 of EXPB1 has three helices, one less than the structure of PcCel45A. However, they are longer than the enzyme's four helices. Two of them (α1 and α3) are in the same position. The β3 sheet of PcCel45A is larger than the β3 of EXPB1. Two of the three disulfide bridges found in EXPB1 are conserved in PcCel45A. The bridge between Cys42 and Cys70 in EXPB1 overlaps with the Cys33 and Cys61 bridge in PcCel45A. Both are located in the β4 sheet (EXPBl) and β4' sheet (PcCel45A). The other disulfide bridge, which is located in exactly the same position in both enzymes, occurs between Cys73 and Cys140 (EXPB1) and Cys64 and Cys149 (PcCel45A). Although the two bridges are in the same position and orientation in relation to the rest of the structure, this bridge links the α4 helix to the β3 sheet in PcCel45A, while in EXPB1 it joins the β3 and β6 strands. All the strands overlap and the α1 and α3 helix of PcCel45A pairs with the α2 and α3 helix of EXPBl, although in PcCel45A they are smaller.

The well-conserved motif known as HFD (hisitidine, phenylalanine and aspartic acid) is part of the conserved active site of GH45[18] and is also found in PcCel45A. It corresponds to residues 119-121 and is found on the β5 strand containing the catalytic aspartate Asp121. In EXPB1 it is found at residues 105-107 and is also on the β5 strand.

The catalytic aspartate of GH45 subfamily A, Asp10, is also not found in EXPB1. However, EXPB1 has an aspartate (Asp95) in the same position as PcCel45A has Asp92 and it has already been proposed that this residue could play a role in the protein's activity.[110]

It has already been described that domain 2 of the expansins has similar characteristics to the type A CBM.[111] The correct functioning of this CBM is not yet known with certainty, but evidence suggests that it binds to the hydrophobic surface of crystalline cellulose with a band of aromatic residues correctly positioned to bind the pyranosidic ring of the substrate. These residues form a planar surface in domain 2 of expansins and their importance has already been proven.[111] As PcCel45A does not have a substrate-binding domain, Trp161, which is located exactly in the position of the CBM domain in expansins, could be related to substrate recognition (Figure 37).

6.6 Structural comparison of PcCel45A and lithium transglycosylase

Using the SPASM[112] computational tool to search the PDB for structures with the Asp-Tyr-Asp triad (relative to Asp92, Tyr25 and Asp121 from PcCel45A), we found a very similar structural motif in a lithium transglycosylase LT) from *E. coli* (EcMltA, PDB 2AE0) and *Neisseria gonorrhoeae* (NgMltA, PDB 2G5D).

The superposition of the structure of PcCel45A with domain 1 of the LTs of the GH102 family revealed significant similarities. The structures overlap with RMSD of 0.6 Â and 0.76 Â for α carbons, respectively. In both cases, the general appearance of the double-Ψ-β-barrel is the same.[46; 113] Furthermore, the overlap of the cellobiose of the PcCel45A-cellobiose complex is very similar to that observed for the cytohexaose in complex with EcMltA[46]. Figure 38 illustrates the similarities in the packaging and positioning of the active site residues.

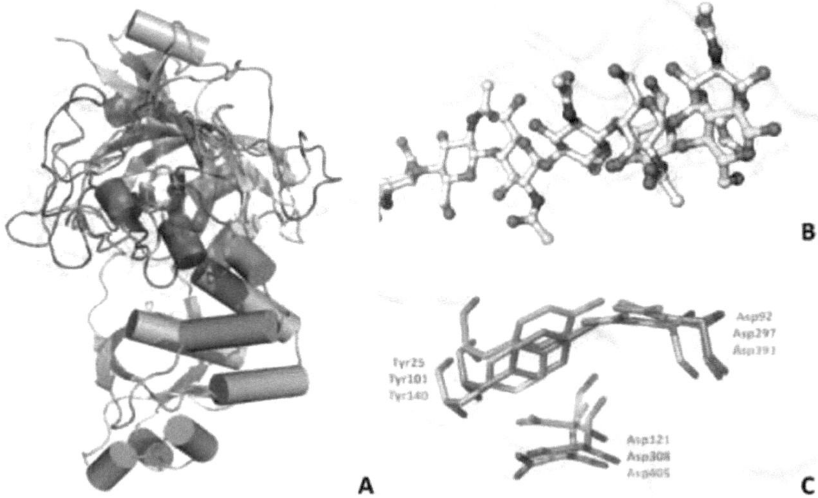

Figure 38 - Comparison between PcCel45A and lithium transglycosylase. A) Overlap of PcCel45A (blue) with domain 1 of the lithium transglycosylase EcMltA (pink). B) Comparison between the position of cellobiose (yellow) and the cytohexaose (gray) of EcMltA shows a similar position in the active site. C) The Asp-Tyr-Asp catalytic triad of PcCel45A (green) superimposed with the residues of EcMltA (pink) and NgMltA (violet) demonstrates that the positioning of these residues is conserved in this enzyme belonging to subfamily C of GH45.

Source: Prepared by the author.

As observed, the structure of PcCel45A is more similar to other classes of proteins (such as β-expansins and LTs) than to the enzymes in the family themselves.

Asp121 is in a hydrophobic environment and its carboxylic group makes a hydrogen bond with the fourth oxygen atom of the cellobiose sugar in the +1 subsite. This fact, together with what was observed for the Hi_EGV enzyme,[56] indicates that this residue takes part in the cleavage reaction.

Analyzing the structure of PcCel45A, we observed that although Asp121 is conserved, there is no other acidic residue in the base position (Asp10 in Hi_EGV), which makes us question the enzyme's mechanism of action. Further analysis of the active site revealed that the enzyme has another aspartic residue, Asp92, at the same distance as Asp10 from Asp121 in Hi_EGV. However, this residue is positioned linearly to Asp121, which would make it difficult to cleave the substrate by inversion of the anomeric carbon.

Despite this, the Asp92 residue is conserved in other proteins such as β-expansins and LTs. Another residue that is conserved in canonical GH45 and LTs and is also found in PcCel45A is Tyr25, which interacts directly with Asp92, forming a bridge between Asp121 and Asp92.

Another interesting feature is that PcCel45A has a long *loop* (*loop* 9) that overlaps with expansin domain 2. At the end of this loop and in the active site cleft is Trp161, which may be involved in substrate binding, since expansin domain 2 also has many aromatic residues involved in substrate binding and PcCel45A has no substrate binding module attached to it.

All this study of the enzyme's active site led us to determine that site-directed mutations in specific residues could explain the enzyme's function. We therefore made mutations in the Asp92, Asp121, Trp161 and Tyr25 residues. All of them were mutated to an alanine.

Chapter 7

Results and discussion

- Production of mutants and analysis of the reaction product

"I think ninety-nine times and discover nothing; I stop thinking," I plunge into deep silence, and the truth reveals itself. "

Albert Einstein

7 Results and Discussion

7.1 Production of mutant DNA and transformation into *A. nidulans*

With the plasmid circularized and transformed into bacteria, we produced enough DNA from the four mutant enzymes for subsequent transformation into *A. nidulans*. Each transformation produces an average of 20 colonies and it is not always that the DNA is incorporated into the DNA of the fungus, which is capable of producing the enzyme.

In this first transformation we found the colony overexpressing the W161A enzyme. We checked the expression of the protein using a 15% SDS gel, as can be seen in figure 39.

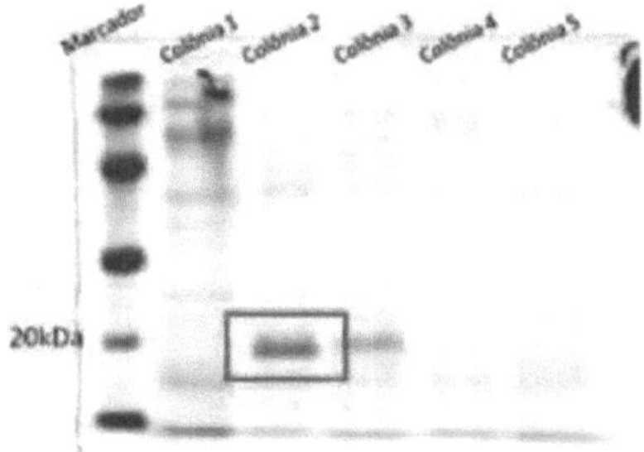

Figure 39 - W161A protein expression test. We can see that the over-expressing colony is colony 2.

Source: Prepared by the author.

Figure 39 shows that protein was expressed in both colony 2 and colony 3. However, as all the samples were concentrated in the same proportion, we can see from the gel that colony 2 expresses more protein than colony 3. It was therefore chosen for further work.

For the Y25A and D121A enzymes, we had to carry out two more transformations in order to find a colony that expressed the protein in large quantities.

For the D92A enzyme, a total of six attempts were made. Figure 40 shows the expression test carried out for this enzyme.

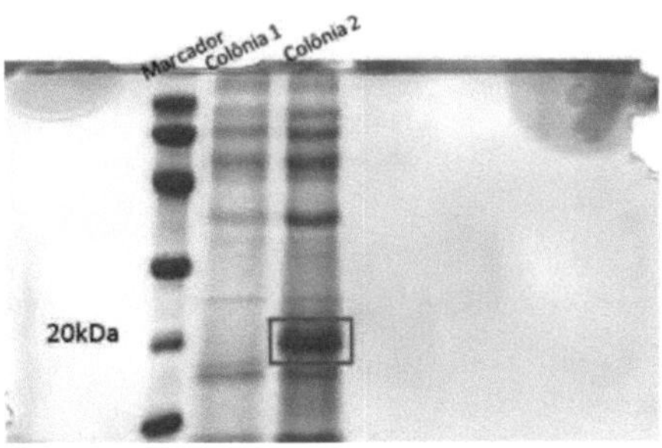

Figure 40 - D92A protein expression test. We can see that the over-expressing colony is colony 2.

Source: Prepared by the author.

Figure 40 shows that the D92A protein is present in colony 2. However, when we compare figure 39 and figure 40, we can clearly see that despite being expressed, the D92A protein has many more contaminants than W161A. Because of this difference, we were unable to complete the purification of this enzyme in a single step.

7.2 Purification

All the enzymes were concentrated in a *Hollow-fiber* cartridge and had their pH increased to 8.5 to facilitate the subsequent purification steps. The concentrated extract was then applied to a *Q-sepharose* column in an *Àkta Purifier* system (GE Healthcare®). After the run was complete, the eluted fractions were applied to an SDS-PAGE gel and by observing the gel we were able to verify the purity of the enzymes produced. The W161A and Y25A proteins from these eluted fractions were used for activity tests. The D121A protein was also used for activity tests and, as we had larger quantities of this enzyme, it was possible to try to crystallize it using commercial sets that presented conditions similar to those under which the native enzyme was crystallized, but we were unsuccessful.

The D92A enzyme required a further chromatographic step for its complete purification. Molecular exclusion chromatography (gel filtration) was used for this. Figure 41 shows the SDS gel of the D92A enzyme after the gel filtration step. This gel was stained with silver nitrate because we obtained a smaller amount than the other enzymes previously purified.

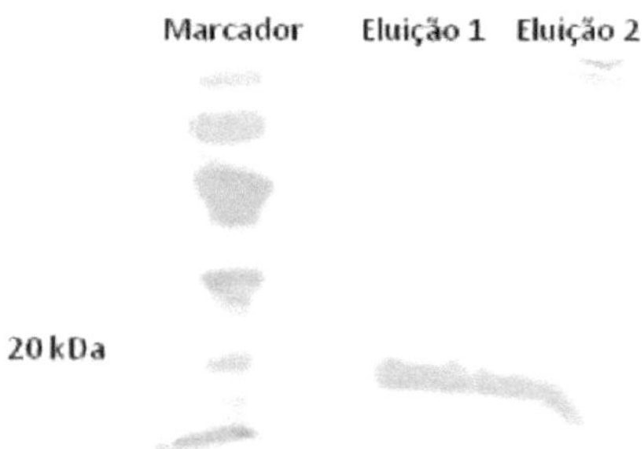

Figure 41 - 15% SDS gel from the purification of the D92A enzyme after molecular exclusion chromatography stained with silver nitrate.

Source: Prepared by the author.

7.3 Mutant activity test

The activity of the mutants was initially tested using the substrate lichenane, as this is the substrate in which the native enzyme showed the highest activity. We used a temperature of 50 °C, even though the optimum temperature for the native enzyme is 70 °C. This is because when we monitored the activity of the native enzyme and mutants at 70 °C, we noticed that the enzyme peaked at 5 minutes and then its specific activity progressively decreased. In addition, the mutations could have destabilized the structure of the enzyme, which would make the mutants less stable at high temperatures.

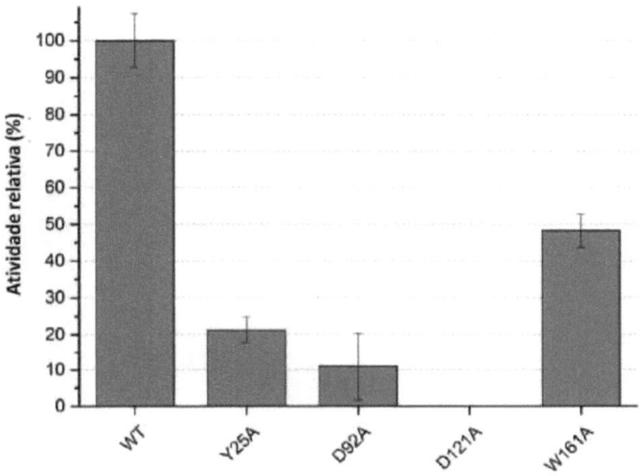

Figure 42 - Relative activity of the native enzyme (WT) and mutants. The results are presented in terms of relative activity. The enzymes were tested at 50 °C for 15 minutes using lichenan as a substrate.

Source: Prepared by the author.

The activity data in Figure 42 suggested that the mutations significantly compromised the activity of the native enzyme. The D121A mutant, as expected, showed no activity, confirming the hypothesis that this residue has an essential function for the enzymatic activity of PcCel45A.

For the W161A mutant, the enzyme lost 50% of its activity, which shows that this residue is important for the enzyme's activity, although the enzyme is still able to remain active, even in its absence. This suggests that this residue has some function related to interaction with the substrate, as we initially hypothesized.

The mutation in Tyr25 (Y25A) also affected the enzyme's activity more severely than W161A, as only 20% of the enzyme's activity remained.

In relation to the D92A mutant, we noticed that there was a great loss of activity (approximately 90% in relation to the native enzyme), however, it was still possible to detect reducing sugars.

HPAE-PAD chromatography was used to confirm the results.

7.4 Product analysis by HPAE-PAD

In order to confirm the effects of the mutations on the enzyme's activity, we used this technique to analyze the reaction products. The result can be seen in Figure 43.

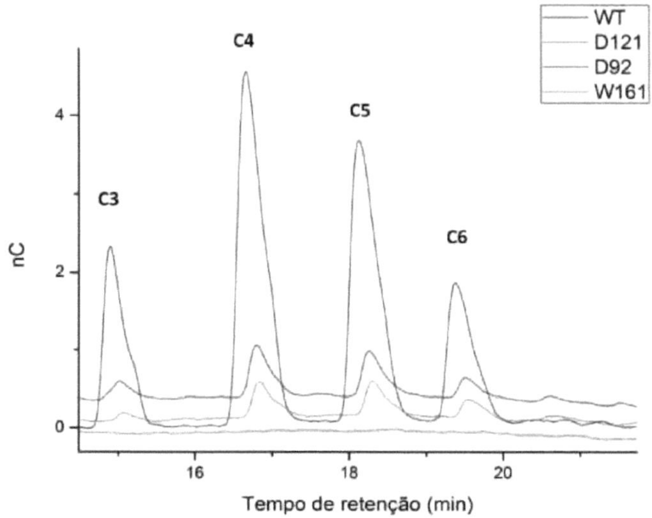

Figure 43 - Celooligosaccharide production by PcCel45A, monitored by HPLC after 24 hours of incubation with PASC. The products of the hydrolysis reaction of the native enzyme (WT) are represented in black and the mutants D121A in red, D92A in blue and W161A in green. The legend above the peaks shows the degree of polymerization of the celooligosaccharides (C3, C4, C5 and C6).

Source: Prepared by the author.

Confirming the results obtained for the native enzyme using TLC, the enzyme showed cellotriose, cellotetraose, cellopentaose and celohexaose as reaction products when PASC was used as a substrate.

We were also able to observe that, with regard to the mutants, the results confirmed the data obtained for activity. The D92A and W161A mutants showed lower activity compared to the native enzyme, while the D121A mutant completely lost its activity.

This result, together with the structure of the enzyme, allows us to better understand its active site and function. As we didn't observe cellobiose as a product, this suggests that the enzyme's subsites must always be filled with molecules larger than two sugars. It seems that the preferred substrates must have between 7 and 9 sugars, since there is a greater release of C4/C5, C4/C4 and C4/C3, respectively.

Compared to the Hi_EGV enzyme, it can also be seen that the latter has the -4, -3, - 2, +1, +2 and +3 sites occupied, again suggesting that the occupation of the -4 to -1 sites is ideal and there is a need to have more than 2 units in the positive site. The difference between them lies in the fact that PcCel45A seems to prefer larger substrates, since it was possible for Hi_EGV to use celohexaose as a substrate, whereas PcCel45A was unable to cleave the same sugar.[56]

This can be explained by analyzing the structures. While PcCel45A has a more elongated (anchor-shaped) cleft, which would easily accommodate a substrate up to C5 in the positive subsites, Hi_EGV is more globular and does not have a cleft long enough to accommodate more than 3 sugar molecules in the positive subsite. This would explain the high specificity for C6 (C3/C3) or C7 (C4/C3) and low activity on longer substrates, since the positive (>C3) and negative (>C4) sites do not interact with the enzyme. Figure 44 shows these differences according to a modeling of the substrates in the enzyme's active site carried out with the Pymol program.[114]

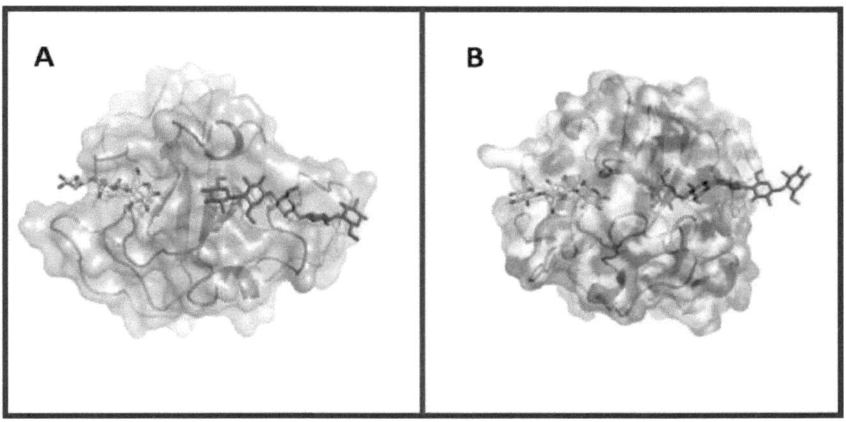

Figure 44 - Comparison between the active site cleft of PcCel45A and Hi_EGV. A) PcCel45A structure with subsites -4, -3 and -2 filled with the "cellotriose" of the Hi_EGV structure and subsites +1, +2 and +3.

filled with a longer polymer overlapping the cellobiose of the PcCel45A-cellobiose complex (in blue). The model has been elongated to show that the +4 and +5 sugars (magenta) would be easily accommodated in the positive site. B) Structure of Hi_EGV with the same sugars shown in A.

Source: Prepared by the author.

Chapter 8

Results and discussion

- Recent results

"For believers, God is at the beginning of things. For

Scientists, at the end of all- reftexdo-. "

Max Planck

8 Recent results

Recently, another research group published the structure of PcCel45A (PcCel45App) and used neutron crystallography to visualize the working mechanism of this enzyme, since this technique allows the hydrogen atoms to be visualized.[115]

It was observed that in the active site of PcCel45A, in a similar position to the proton acceptor residue in Hi_EGV (Asp10), there was an asparagine (Asn99 from PcCel45A), which was proposed by this group to act as a proton acceptor in the inversion reaction, as can be seen in figure 45.

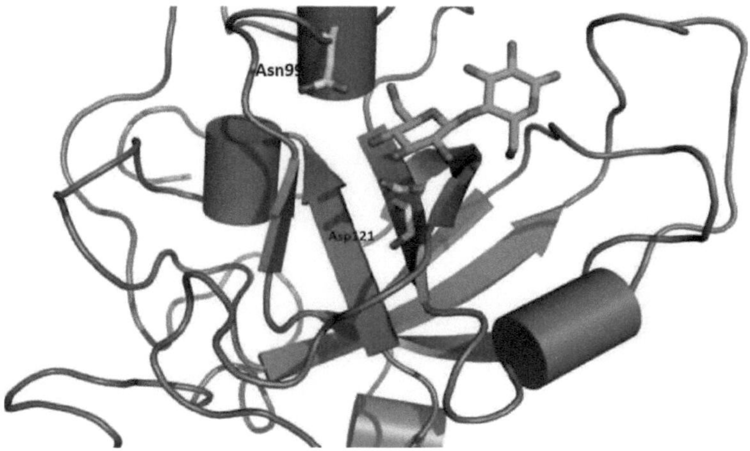

Figure 45 - Active site of PcCel45A showing the position of asparagine (Asn99) compared to the catalytic residue Asp121.

Source: Prepared by the author.

The amino acid asparagine has a carboxyamide group as its side chain, which is why it is considered to be an amide of aspartic acid. It is a neutral, polar and uncharged amino acid at any biologically relevant pH,[116] so initially we didn't consider that it could be part of the catalysis reaction.

Using X-ray crystallography and neutron crystallography techniques, it was possible to observe that the side chain of the asparagine residue takes on an acidic imide form (HO-CX =NH) and not the typical amide form (O=CX-NHz) in the structure. The estimated pKa value for the protonation/deprotonation of the nitrogen of the imide form is estimated at 4.5 to 7.5, while for the oxygen it is around 0. This information supports the idea that the nitrogen of asparagine may be the catalytic base of PcCel45A, although it does not clearly explain how the oxygen atom of asparagine is protonated.

Analysis of the map by neutron crystallography showed that the nitrogen atoms in the vicinity of the amide of Cys96 and the oxygen atoms in the side and main chain of Asn99 were connected. Intramolecular proton transfer in hydrogen bonds between the carbonyl and amide groups of small molecules often occurs if the positions of the nitrogen and oxygen are suitable. This possibility supports the idea that proton transfer from the amide group of Cys96 to the carbonyl of the side group of Asn99 can occur and this would be consistent with the stabilization of the acidic-imide form of Asn99 by multiple tautomerizations on the enzyme surface.

These results were only possible due to neutron crystallography, because when comparing with X-ray crystallography data, even obtaining a resolution of 0.65 Â, it was not possible to observe the hydrogens in the protein molecule.[115]

Chapter 9

Conclusion and outlook

"And the flower said to the little prince, 'I need to bear one or three larvae to meet the pumpkins. "

Antoine Saint-Exupéry

9 Conclusions and prospects

This work arose from the need to obtain more information about the 45 family of glycoside hydrolases. To this end, our efforts were focused on the endoglucanase V enzyme from *Phanerochaete chrysosporium* (PcCel45A), which, despite having been biochemically characterized by another group, had some incomplete information and its structural model was not available. This enzyme was chosen because it belongs to subfamily C, a subfamily with no structural information available.

To this end, our objectives included the biochemical characterization of the PcCel45A enzyme, as well as its structural elucidation. To produce the enzyme, we used the LIC system for cloning, as well as heterologous expression in a filamentous fungus (*Aspergillus nidulans*).

Analysis of the enzyme's biochemical profile confirmed its acidophilic characteristics (optimum pH 4) and a high optimum temperature (70 °C). In terms of substrate preference, PcCel45A showed the highest activity for the substrate lichenan, followed by filter paper, β-glucan, galactomannan and CMC. It was not possible to detect the production of celooligosaccharides when the substrate contains β1,3 bonds, however, when using PASC as a substrate, we were able to detect the production of celooligosaccharides with polymerization degrees 3, 4 and 5 when analyzing the products by TLC. For a more sensitive detection of the hydrolysis reaction products, we used the HPAE-PAD technique, in which it was possible to detect celooligosaccharides with 3, 4, 5 and 6 sugars, with the highest peaks relating to sugars with 4 and 5 units.

By examining the active site of PcCel45A, it can be seen that the preference for larger substrates may be due to the long cleft and anchor shape of the active site of this enzyme. Models generated by Pymol have shown that the active site of this enzyme is able to accommodate larger substrates (with up to 10 sugars) than that of other enzymes in the family. This observation may explain why we don't find small celooligosaccharides when we use substrates with β1,3 bonds.

Structure determination revealed very interesting and unique characteristics of the enzyme studied. Its main feature is a double-Ψ-β-barrel folding, which is conserved in other enzymes. Despite this similarity, the overall shape of PcCel45A is very different from other GH45s. The PcCel45A enzyme has a more elongated cleft, which seems to better accommodate the substrate in the active site, by means of aromatic residues, while the other enzymes have a smaller cleft and use *loops* that surround the substrate.

Another interesting fact is that PcCel45A is more similar to members belonging to protein classes such as β-expansins and LTs than to the GH family 45 enzymes themselves. This can be seen by looking at the RMSD values obtained when the structures are superimposed.

The structural model obtained for PcCel45A also showed the absence of the Asp10 residue, which is the hydrogen acceptor in the water molecule that allows the substrate to be cleaved by the inversion reaction, which is usual for the other GH45s.

PcCel45A has been shown to have another aspartate, Asp92, at the same distance as Asp121 (8 Â). This residue is conserved in other protein classes such as β-expansins and LTs. However, this residue is in the active site, positioned linearly to the Asp121 residue, which would make it difficult to cleave the substrate.

Still looking at the active site of PcCel45A, we noticed that there is a Tyr25 that is conserved in other

GH45 and similar classes. This Tyr25 is in direct contact with Asp92 and close to Asp121, which led us to assume that it might play a role in the enzyme's activity.

Another interesting residue is Trp161, which is positioned in *loop* 9 of the enzyme, in the region that forms the end of the surface, a position in which, when we superimpose PcCel45A and β-expansins, domain 2 of the latter is found. It is speculated that this domain functions as a type A CBM, which binds the substrate by means of aromatic residues.

All these observations about the enzyme's active site led us to make site-directed mutations in these residues. Thus, we produced the mutants D121A, D92A, Y25A and W161A.

The activity tests showed that the W161A mutant lost 50% of its activity. Y25A lost 80%, while the D92A mutant lost an average of 90% of its activity. As expected, the D121A mutant lost all its activity, confirming its catalytic importance.

To confirm these results, we analyzed the reaction product for each enzyme. Again, the results for D121A confirmed the annihilation of this mutant's activity, while W161A and D92A lost activity to a similar extent. We did not obtain enough Y25A for this test. These results showed that, indeed, these residues have a functional importance in the activity of the enzyme.

Recently, a study published on the same enzyme found that the catalytic base was an asparagine, Asn99 in the structure of PcCel45A, which in its imidic state was capable of being the acceptor in the catalysis reaction. This observation was only possible thanks to neutron crystallography, which is able to visualize the hydrogens in the structure.

We believe that this work has shown that although much has been studied about glycoside hydrolases, there are still some gaps that need to be filled. Family 45 is an example of this, due to the limited information available. This work has contributed to expanding our knowledge of this family, demonstrating through the structural model, the similarities of this family with the β-expansins and LTs, which had previously only been discussed through amino acid sequence analysis.

The prospects of this work are to seek the structural elucidation of more enzymes from the GH45 family, focusing on subfamilies B and C, which still need structural information. Through biochemical study and analysis of the structural model, it could be confirmed whether more enzymes have this atypical mechanism of action.

REFERENCES

1 HIMMEL, M. E.; RUTH, M. F.; WYMAN, C. E. Cellulase for commodity products from cellulosic biomass. **Current Opinion Biotechnology,** v. 10, n. 4, p. 358-64, Aug. 1999.

2 LYND, L. R. et al. Microbial cellulose utilization: fundamentals and biotechnology. **Microbiology and Molecular Biology Reviews,** v. 66, n. 3, p. 506-577, Sept. 2002.

3 SADDLER J., L. A.; KADOLPH S. **Textiles**. 7th ed. New Jersey: Prentice Hall, 2001. 427p.

4 RAVINDRAN, R.; JAISWAL, A. K. A comprehensive review on pre-treatment strategy for lignocellulosic food industry waste: challenges and opportunities. **Bioresource Technology**, Aug. 2015. Available at: < http://dx.doi.org/10.1016/j.biortech.2015.07.106 > Accessed on: Oct. 2015.

5 CHAWLA, P. R. et al. Microbial cellulose: fermentative production and applications. **Food Technology and Biotechnology,** v. 47, n. 2, p. 107-124, 2015.

6 STANGHERLIN, S. **Advantages and disadvantages of enzymatic and chemical hydrolysis in the production of bioethanol from sugarcane bagasse**. Available at:< http://pesquisa.unifacs.br/wp-content/uploads/2015/04/F%C3%A1bio-Cirqueira-da- Silva.pdf>. Accessed on: Oct. 2015.

7 YANG, B. et al. Enzymatic hydrolysis of cellulosic biomass. **Biofuels.** v. 02, n. 4, p. 421449, 2014

8 WYMAN, C. E. Biomass ethanol: technical progress, opportunities, and commercial challenges. **Annual Review of Energy and the Environment,** v. 24, p. 189-226, Nov. 1999. doi: 10.1146/annurev.energy.24.1.189.

9 ZHANG, Y. H.; LYND, L. R. Toward an aggregated understanding of enzymatic hydrolysis of cellulose: noncomplexed cellulase systems. **Biotechnology and Bioengineering,** v. 88, n.

7, p. 797-824, Dec. 2004.

10 CANTAREL, B. L. et al. The Carbohydrate-Active EnZymes database (CAZy): an expert resource for Glycogenomics. **Nucleic Acids Research**, v.37, p. D233-8, Jan. 2009. doi: 10.1093/nar/gkt1178.

11 HENRISSAT, B.; BAIROCH, A. New families in the classification of glycosyl hydrolases based on amino acid sequence similarities. **Biochemical Journal,** v. 293, n. 3, p. 781-788, Aug. 1993.

12 IGARASHI, K. et al. Characterization of endoglucanase belonging to new subfamily of glycoside hydrolase family 45 from the basidiomycete *Phanerochaete chrysosporium*.

Applied and Environmental Microbiology, v. 74, n. 18, Aug. 2008. doi:

10.1128/AEM.00812-08.

13 GAO, D. et al. A critical review of the application of white rot fungus to environmental pollution control. **Critical Reviews in Biotechnology,** v. 30, n. 1, p. 70-7, Mar. 2010.

14 SZABÓ, I. J.; JOHANSSON, G.; PETTERSSON, G. Optimized cellulase production by *Phanerochaete chrysosporium*: control of catabolite repression by fed-batch cultivation. **Journal of Biotechnology,** v. 48, n. 3, p. 221-230, July 1996.

15 KHALIL, A. I. Production and characterization of cellulolytic and xylanolytic enzymes from the ligninolytic white-rot fungus *Phanerochaete chrysosporium* grown on sugarcane bagasse. **World Journal of Microbiology and Biotechnology,** v. 18, n. 8, p. 753-759, Nov. 2002.

16 TAIZ, L.; ZEIGER, E. **Plant physiology**. 5th ed. Sunderland, MA: Sinauer Associates, 2010. 782 p.

17 DOBLIN, M. S.; PETTOLINO, F.; BACIC, A. Evans review: plant cell walls: the skeleton of the plant world. **Functional Plant Biology,** v. 37, n. 5, p. 357-381. Apr. 2010.

18 COSGROVE, D. Loosening of plant cell walls by expansins. **Nature,** v. 407, p. 321 - 326, 2000. doi:10.1038/35030000

19 SZYMANSKI, D. B.; COSGROVE, D. J. Dynamic coordination of cytoskeletal and cell wall systems during plant cell morphogenesis. **Current Biology,** v. 19, n. 17, p. 800-811, Sept. 2009.

20 CARPITA, N. C. et al. Cell wall architecture of the elongating maize coleoptile. **Plant Physiology,** v. 127, n. 2, p. 551-565, Jan. 2001.

21 RUBIN, E. M. Genomics of cellulosic biofuels. **Nature,** v. 454, n. 7206, p. 841-845, Aug. 2008.

22 VORAGEN, A. J. et al. Pectin, a versatile polysaccharide present in plant cell walls. **Structural Chemistry,** v. 20, n. 2, p. 263-275, Apr. 2009.

23 MOHNEN, D. Pectin structure and biosynthesis. **Current Opinion in Plant Biology,** v.

11, n. 3, p. 266-77, June 2008.

24 WILLATS, W. G. T.; KNOX, J. P.; MIKKELSEN, J. D. Pectin: new insights into an old polymer are starting to gel. **Trends in Food Science & Technology,** v. 17, n. 3, p. 97-104, Mar. 2006.

25 SYED, K.; YADAV, J. S. P450 monooxygenases (P450ome) of the model white rot fungus *Phanerochaete chrysosporium*. **Critical Reviews in Microbiology,** v. 38, n. 4, p. 33963, Nov. 2012.

26 WHITING, P.; GORING, D. A. I. Chemical characterization of tissue fractions from the middle lamella and secondary wall of black spruce tracheids. **Wood Science and Technology,** v. 16, n. 4, p. 261-267, Dec. 1982

27 ROBYT, J. F. **Essentials of carbohydrate chemistry**. New York: Springer Verlag, 1998. 399 p.

28 DAVIN, L. B.; LEWIS, N. G. Lignin primary structures and dirigent sites. **Current Opinion in Biotechnology,** v. 16, n. 4, p. 407-15, Aug. 2005.

29 EMSLEY, A. M.; STEVENS, G. C. Kinetics and mechanisms of the low-temperature degradation of cellulose. **Cellulose,** v. 1, n. 1, p. 26-56, Mar. 1994.

30 PAWCENIS, D. et al. Towards determination of absolute molar mass of cellulose polymer by size exclusion chromatography with mulitple angle laser light scattering detection. **Journal of Chromatography A**, v. 1409, p.53-59, June, 2015.

doi:10.1016/j.chroma.2015.06.042

31 BACIC, E. A. **The biochemistry of plants:** structure and function of plant cell wall. San Diego: Academic Pres Inc., 1988. 529p.

32 RAYLE, D. L.; CLELAND, R. E. The acid growth theory of auxin-induced cell elongation is alive and well. **Plant Physiology,** v. 99, n. 4, p. 1271-4, Aug. 1992.

33 LIPCHINSKY, A. How do expansins control plant growth? a model for cell wall loosening via defect migration in cellulose microfibrils. **Acta Physiologiae Plantarum,** v. 35, n. 12, p. 3277-3284, Dec. 2013.

34 COSGROVE, D. Characterization of long-term extension of isolated cell walls from growing cucumber hypocotyls. **Planta,** v. 177, n.1, p. 121 - 130, 1989.

35 MCQUEEN-MASON, S.; DURACHKO, D.; COSGROVE, D. Two endogenous proteins that induce cell wall expansion in plants. **Plant Cell,** v. 4, n.11, p. 1425 - 1433, 1992.

36 CHO, H.T.; COSGROVE, D. J. Altered expression of expansin modulates leaf growth and pedicel abscission in Arabidopsis thaliana. **PNAS,** v. 97, n. 17, p. 9783-9788, Aug. 2000.

37 WU, Y. et al. Modification of expansin transcript levels in the maize primary root at low water potentials. **Plant Physiology,** v. 126, n. 4, p. 1471-9, Aug. 2001.

38 NAUMOFF, D. G. Hierarchical classification of glycoside hydrolases. **Biochemistry,** v. 76, n. 6, p. 622-35, June 2011.

39 LOMBARD, V. et al. The Carbohydrate-active enzymes database (CAZy) in

2013. **Nucleic Acids Research, v.** 42, p. D490-5, 2014. doi: 10.1093/nar/gkt1178

40 B HENRISSAT, M. C.; TOMME,L.P.; LEMESLE,J.; MORNON, P.. Cellulase families revealed by hydrophobic cluster analysis. **Gene,** v. 81, n. 1, p. 83-95, 1989.

41 HENRISSAT, B. A classification of glycosyl hydrolases based on amino acid sequence similarities. **Biochemical Journal,** v. 280, n. 2, p. 309-16, Dec. 1991.

42 DAVIES, G.; HENRISSAT, B. Structures and mechanisms of glycosyl hydrolases. **Structure,** v. 3, n. 9, p. 853-859, Sept. 1995.

43 MCCARTER, J. D.; STEPHEN WITHERS, G. Mechanisms of enzymatic glycoside hydrolysis. **Current Opinion in Structural Biology,** v. 4, n. 6, p. 885-892, 1994.

44 WITHERS, S. G.; AEBERSOLD, R. Approaches to labeling and identification of active site residues in glycosidases. **Protein Science,** v. 4, n. 3, p. 361-72, Mar. 1995.

45 JUERS, D. H.; HUBER, R. E.; MATTHEWS, B. W. Structural comparisons of TIM barrel proteins suggest functional and evolutionary relationships between beta-galactosidase and other glycohydrolases. **Protein Science,** v. 8, n. 1, p. 122-36, Jan. 1999.

46 VAN STRAATEN, K. E. et al. Structure of Escherichia coli Lytic transglycosylase MltA with bound chitohexaose: implications for peptidoglycan binding and cleavage. **Journal of Biological Chemistry**, v. 282, n. 29, p. 21197-205, 2007.

47 FELIX, C. R.; LJUNGDHAL, L. G. The cellulosome: the exocellular organelle of clostridium. **Annual Review of Microbiology**, v. 47, p. 791-819, 1993.

doi: 10.1146/annurev.mi.47.100193.004043

48 HARREITHER, W. et al. Cellobiose Dehydrogenase from the Ligninolytic Basidiomycete Ceriporiopsis subvermispora. **Applied and Environmental Microbiology,** v. 75, n. 9, p.

2750-2757, May 2009.

49 KIRK, K.; CULLEN, D. **Environmentally friendly technologies for the pulp and paper industry**: enzymology and molecular genetics of wood degradation by white-rot fungi. New York: John Wiley and Sons, 1998. 577p.

50 MORGENSTERN, I.; POWLOWSKI, J.; TSANG, A. Fungal cellulose degradation by oxidative enzymes: from dysfunctional GH61 family to powerful lytic polysaccharide monooxygenase family. **Briefings Functional Genomics,** v. 13, n. 6, p. 471-81, Nov. 2014.

51 BHATTACHARYA, A. S.; BHATTACHARYA, A.; PLETSCHKE, B. I. Synergism of fungal and bacterial cellulases and hemicellulases: a novel perspective for enhanced bioethanol production. **Biotechnology Letters,** v. 37, n. 6, p. 1117-29, June 2015.

52 RABINOVICH, M. L.; MELNICK, M. S.; BOLOBOVA, A. V. The structure and mechanism of action of cellulolytic enzymes. **Biochemistry,** v. 67, n. 8, p. 850-871, Aug. 2002.

53 YENNAMALLI, R. M. et al. Endoglucanases: insights into thermostability for biofuel applications. **Biotechnology for Biofuels,** v. 6, n. 1, p. 136, Sept. 2013.

54 SEGATO, F. et al. Genomics review of holocellulose deconstruction by aspergilli. **Microbiology and Molecular Biology Reviews,** v. 78, n. 4, p. 588-613, Dec. 2014.

55 VLASENKO, E. et al. Substrate specificity of family 5, 6, 7, 9, 12, and 45 endoglucanases. **Bioresource Technology,** v. 101, n. 7, p. 2405-11, Apr. 2010.

56 DAVIES, G. et al. Structures of oligosaccharide-bound forms of the endoglucanase V from Humicola insolens at 1.9 A resolution. **Biochemistry,** v. 34, n. 49, p. 16210 - 16220, 1995.

57 DAVIES, G. J. et al. Mapping the conformational itinerary of beta-glycosidases by X-ray crystallography. **Biochemical Society Transactions,** v. 31, n. 3, p. 523-527, June 2003.

58 COUTURIER, M. et al. A thermostable GH45 endoglucanase from yeast: impact of its atypical multimodularity on activity. **Microbial Cell Factories,** v. 10, n. 1, p. 103, 2011.

59 PARAMESWARAN, B.; RAVEENDRAN, S.; ASHOK, P. Provided for hydrolysis of

Iignocellulosic biomass for bioethanol production. **Biofuels**, 2011. Available at: < http://www.researchgate.net/publication/227414108_Provided_for_Hydrolysis_of_Lignocellulosic_Biomass_for_Bioethanol_Production> Accessed on: 02 nov. 2015.

60 HORN, S. J. et al. Novel enzymes for the degradation of cellulose. **Biotechnology for Biofuels,** v. 5, n. 1, p. 45, July 2012.

61 FUSHINOBU, S. Metalloproteins: a new face for biomass breakdown. **Nature Chemical Biology,** v. 10, n. 2, p. 88-89, Dec. 2013.

62 DAVIES, G. J.; WILSON, K. S.; HENRISSAT, B. Nomenclature for sugar-binding subsites in glycosyl hydrolases. **Biochemical Journal,** v. 321, n. 2, p. 557-9, Jan. 1997.

63 DAVIES, G. J. et al. Structure determination and refinement of the Humicola insolens endoglucanase V at 1.5 A resolution. **Acta Crystallographica Section D:** biological crystallography, v.52, n.1, p.7-17, 1996.

64 PAYNE, C. M. et al. Fungal cellulases. **Chemical Reviews,** v. 115, n. 3, p. 1308-1448, Jan. 2015.

65 CASTILLO, R. M. et al. A six-stranded double-psi β barrel is shared by several protein superfamilies. **Structure,** v. 7, n. 2, p. 227-236, Feb. 1999.

66 SCHOU, C. et al. Stereochemistry, specificity and kinetics of the hydrolysis of reduced cellodextrins by nine cellulases. **European Journal Biochemistry,** v. 217, n. 3, p. 947 - 953, Nov. 1993.

67 HIRVONEN, M.; PAPAGEORGIOU, A. C. Crystal structure of a family 45 endoglucanase from melanocarpus albomyces: mechanistic implications based on the free and cellobiose-bound forms. **Journal of Molecular Biology,** v. 329, n. 3, p. 403-410, June 2003.

67 GUERRIERO, G. et al. Destructuring plant biomass: focus on fungal and extremophilic cell wall hydrolases. **Plant Science,** v. 234, p. 180-193, May 2015.

doi:10.1016/j.plantsci.2015.02.010.

69 MARTINEZ, D. et al. Genome sequence of the lignocellulose degrading fungus *Phanerochaete chrysosporium* strain RP78. **Nature Biotechnology,** v. 22, n. 6, p. 695-700, June 2004.

70 ASGHER, M. et al. Recent developments in biodegradation of industrial pollutants by white rot fungi and their enzyme system. **Biodegradation,** v. 19, n. 6, p. 771-83, Nov. 2008.

71 ZACCHI, L. et al. Metabolism of cellulose by *Phanerochaete chrysosporium* in continuously agitated culture is associated with enhanced production of lignin peroxidase. **Journal of Biotechnology,** v. 78, n. 2, p. 185-192, Oct. 2000.

72 ADAV, S. S.; RAVINDRAN, A.; SZE, S. K. Quantitative proteomic analysis of lignocellulolytic enzymes by *Phanerochaete chrysosporium* on different lignocellulosic biomass. **Journal of Proteomics,** v. 75, n. 5, p. 1493-1504, Feb. 2012.

73 KIRK, T. K.; FARRELL, R. L. Enzymatic "combustion": the microbial degradation of lignin. **Annual Review of Microbiology,** v. 41, p. 465-505, 1987.

doi: 10.1146/annurev.mi.41.100187.002341

74 KAVANAGH, K. **Fungi:** biology and applications. 2nd ed. New Jersey: Wiley, 2011. 376p.

75 BALDRIAN, P.; VALASKOVA, V. Degradation of cellulose by basidiomycetous fungi. **FEMS Microbiology Reviews,** v. 32, n. 3, p. 501-21, May 2008.

76 LIN, W. L.et al. Kinetics of cell growth and heterologous glucoamylase production in recombinant *Aspergillus nidulans*. **Biotechnology Bioengeneering,** v. 41, n. 2, p. 273-9, Jan. 1993.

77 MEYER, V.; WU, B.; RAM, A. F. Aspergillus as a multi-purpose cell factory: current status and perspectives. **Biotechnology Letters,** v. 33, n. 3, p. 469-76, Mar. 2011.

78 PETERSEN, N. T.; BRUNAK, S.; NIELSEN, H. SignalP 4.0: discriminating signal peptides from transmembrane regions. **Nature Methods,** v. 8, p. 785-786, 2011.

doi:10.1038/nmeth.1701

79 EMANUELSON, O.; NIELSEN, H.; Predicting subcellular localization of proteins based on their N-terminal amino acid sequence. **Journal of Molecular Biology.** v. 300, n. 4, p. 1005-1016,

2000.

80 RACC - NIH. **Laboratory for structural genomics and proteomics**. Available at: <http://nihserver.mbi.ucla.edu/RACC/> Accessed on: May 2012.

81 NETNGLYC. **Center for biological sequence analysis**. Available at:

<http ://www. cb s. dtu. dk/service s/N etN Glyc/> Accessed on: May 2012

82 ASLANIDIS, C.; DE JONG, P. J. Ligation-independent cloning of PCR products (LIC- PCR). **Nucleic Acids Reviews,** v. 18, n. 20, p. 6069-74, Oct. 1990.

83 SEGATO, F.; GONÇALVES, A.T.; LUCAS, C.R.; SQUINA, F.; PRADE, R. High-yield secretion of multiple client proteins in Aspergillus. **Enzyme and Microbial Technology,** v. 51, n. 2, p. 100-6, July 2012.

84 SAMBROOK, J.; RUSSEL, D. W. **Molecular cloning, a laboratory manual**. U.S.A: Cold Spring Harbor Laboratory Press, 2001.

85 RABHI, I. et al. A novel simple and rapid PCR-based site-directed mutagenesis method. **Molecular Biotechnology,** v. 26, n. 1, p. 27-34, Jan. 2004.

86 BARRATT, R.W.; JOHNSON, G.B.; OGATA, W.N. Wild-type and mutant stocks of *Aspergillus nidulans*. **Genetics**, v. 52, n. 1, p. 233-246, 1965.

87 GE Healthecare LifeScience. Available at: < https://www.gelifesciences.com> Accessed: May 2013

88 GASTEIGER E., H. C.; GATTIKER A., DUVAUD, S.; WILKINS, M.R.; APPEL, R.D.; BAIROCH, A. **The proteomics protocols handbook**: protein identification and analysis tools on the expasy server. Totowa, NJ: Humana Press, 2005. p. 571-607.

89 BAILEY, M. A note on the use of dinitrosalicylic acid for determining the products of enzymatic reactions. **Applied Microbiology and Biotechnology,** v. 29, n. 5, p. 494-496, Nov. 1988.

90 THERMOSCIENTIFIC. **Thermo fisher scientific**.Available at:< http://www.thermoscientific.com/en/home.html> Accessed: Oct. 2015.

91 DRENTH, J. **Principles of protein x-ray crystallography.** 3rd ed. Groningen: Springer. 2007.345p.

92 LEUNISSEN, M. **An essay on several aspects of protein crystallization research.**

2001. Available at:< http://people.ds.cam.ac.uk/ml527/publications/assets/leunissen-literatureresearch.pdf >. Accessed on: October 23, 2015,

93 MCPHERSON, A. Introduction to protein crystallization. **Methods,** v. 34, n. 3, p. 254265, Nov. 2004.

94 KRAUSS, I. R. et al. An overview of biological macromolecule crystallization. **International Journal of Molecular Sciences,** v. 14, n. 6, p. 11643-11691, May 2013.

95 CSIRO. **Collaborative Crystallization Center.** Available at:<www.csiro.au/C3) >. Accessed: Oct. 2015.

96 KABSCH, W. XDS. **Acta Crystallographica Section D,** v. 66, n. 2, p. 125-32, 2010.

97 BRUKER. **APEX2.**Wisconsin, USA. 2007.

98 MCCOY, A. J. et al. Phaser crystallographic software. **Journal of Applied Crystallography,** v. 40, n. 4, p. 658-674, July 2007.

99 SODING, J.; BIEGERT, A.; LUPAS, A. N. The HHpred interactive server for protein homology detection and structure prediction. **Nucleic Acids Research,** v. 33, p. 244-8, July 2005. doi: 10.1093/nar/gki408

100 ADAMS, P. D. et al. PHENIX: a comprehensive Python-based system for macromolecular structure solution. **Acta Crystallographica Section D,** v. 66, n. 2, p. 213221, 2010.

101 EMSLEY, P. et al. Features and development of Coot. **Acta Crystallographica Section D:** biology crystallography, v. 66, n. 4, p. 486-501, 2010.

102 AFONINE, P. V. et al. Towards automated crystallographic structure refinement with phenix.refine. **Acta Crystallographica Section D:** biology crystallography, v. 68, n. 4, p. 352-67, Apr. 2012.

103 CHEN, V. B. et al. MolProbity: all-atom structure validation for macromolecular crystallography. **Acta Crystallographica Section D:** biology crystallography, v. 66, n. 1, p. 12-21, Jan 2010.

104 SCHAGGER, H.; VON JAGOW, G. Tricine-sodium dodecyl sulfate-polyacrylamide gel electrophoresis for the separation of proteins in the range from 1 to 100 kDa. **Analytical Biochemistry,** v. 166, n. 2, p. 368-79, Nov. 1987.

105 WONGANU, B. et al. Cloning, expression and characterization of a thermotolerant endoglucanase from Syncephalastrum racemosum (BCC18080) in Pichia pastoris. **Protein Expression and Purification.** v. 58, n. 1, p.78-86, 2008.

106 LIU, G.; WEI, X.; QIN, Y.; QU, U. Characterization of the endoglucanase and glucomannanase activities of a glycoside hydrolase family 45 protein from Penicillium decumbens 114-2. **Journal of**

General and Applied Microbiology. v. 56, n. 3, p. 223-9, June 2010.

107 KOGA, J. et al. Purification and characterization of a new family 45 endoglucanase, STCE1, from Staphylotrichum coccosporum and its overproduction in Humicola insolens. **Applied and Environmental Microbiology.** v. 74, n. 33, p. 4210-7, July 2008.

108 KARLSSON, J. et al. Enzymatic properties of the low molecular mass endoglucanases Cel12A (EG III) and Cel45A (EG V) of Trichoderma reesei. **Journal of Biotechnology.** v. 99, n. 1, p. 63-78, Oct. 2002.

109 KADOWAKI, M. A. et al. Functional characterization and low-resolution structure of an endoglucanase cel45a from the filamentous fungus neurospora crassa or74a: thermostable enzyme with high activity toward lichenan and beta-glucan. **Molecular Biotechnology,** v. 57, n. 6, p. 574-88, June 2015.

110 YENNAWAR, N. H. et al. Crystal structure and activities of EXPB1 (Zea m 1), a β- expansin and group-1 pollen allergen from maize. **Proceedings of the National Academy of Sciences,** v. 103, n. 40, p. 14664-14671, 2006.

111 GEORGELIS, N.; YENNAWAR, N. H.; COSGROVE, D. J. Structural basis for entropy- driven cellulose binding by a type-A cellulose-binding module (CBM) and bacterial expansin. **Proceedings of the National Academy of Sciences of the United States of America,** v. 109, n. 37, p. 14830-14835, Aug. 2012.

112 KLEYWEGT, G. J. Recognition of spatial motifs in protein structures. **Journal of Molecular Biology**, v. 285, n. 4, p. 1887-97, 1999.

113 POWELL, A. J. et al. Crystal structures of the lytic transglycosylase MltA from N.gonorrhoeae and E.coli: insights into interdomain movements and substrate binding. **Journal of Molecular Biology**, v. 359, n. 1, p. 122-36, 2006.

114 The PyMOL Molecular Graphics System: version 1.7.4. Available at: < https://www.pymol.org/citing>. Accessed on: Oct. 2015.

115 NAKAMURA, A. et al. "Newton's cradle" proton relay with amide-imidic acid tautomerization in inverting cellulase visualized by neutron crystallography. **Science Advances,** v. 1, n. 7, Aug. 2015. doi: 10.1126/sciadv.1500263

116 TUTORVISTA. Available at:

<http://chemistry.tutorvista.com/biochemistry/asparagine.html>. Accessed Oct. 2015.

MIX
Papier aus verantwortungsvollen Quellen
Paper from responsible sources
FSC® C105338

Printed by Books on Demand GmbH, Norderstedt / Germany